Four-Dimensional Paper Constructions
After Möbius, Klein and Boy

SERIES ON KNOTS AND EVERYTHING

ISSN: 0219-9769

Editor-in-charge: Louis H. Kauffman *(Univ. of Illinois, Chicago)*

The Series on Knots and Everything: is a book series polarized around the theory of knots. Volume 1 in the series is Louis H Kauffman's Knots and Physics.

One purpose of this series is to continue the exploration of many of the themes indicated in Volume 1. These themes reach out beyond knot theory into physics, mathematics, logic, linguistics, philosophy, biology and practical experience. All of these outreaches have relations with knot theory when knot theory is regarded as a pivot or meeting place for apparently separate ideas. Knots act as such a pivotal place. We do not fully understand why this is so. The series represents stages in the exploration of this nexus.

Details of the titles in this series to date give a picture of the enterprise.

Published:

More information on this series can also be found at http://www.worldscientific.com/series/skae

Series on Knots and Everything — Vol. 78

Four-Dimensional Paper Constructions
After Möbius, Klein and Boy

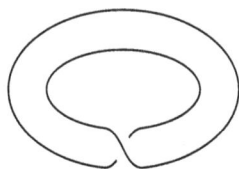

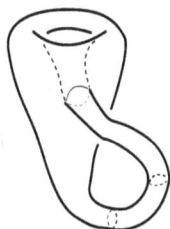

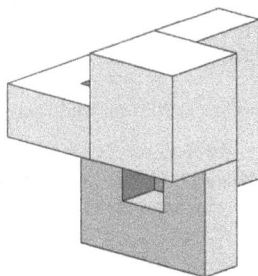

Eiji Ogasa
Meiji Gakuin University, Japan

World Scientific

NEW JERSEY · LONDON · SINGAPORE · BEIJING · SHANGHAI · HONG KONG · TAIPEI · CHENNAI

Published by

World Scientific Publishing Co. Pte. Ltd.

5 Toh Tuck Link, Singapore 596224

USA office: 27 Warren Street, Suite 401-402, Hackensack, NJ 07601

UK office: 57 Shelton Street, Covent Garden, London WC2H 9HE

Library of Congress Control Number: 2024042791

British Library Cataloguing-in-Publication Data
A catalogue record for this book is available from the British Library.

Series on Knots and Everything — Vol. 78
FOUR-DIMENSIONAL PAPER CONSTRUCTIONS AFTER MÖBIUS, KLEIN AND BOY

ISBN 978-981-98-0179-4 (hardcover)
ISBN 978-981-98-0180-0 (ebook for institutions)
ISBN 978-981-98-0181-7 (ebook for individuals)

For any available supplementary material, please visit
https://www.worldscientific.com/worldscibooks/10.1142/14069#t=suppl

Desk Editors: Murali Appadurai/Angeline Husni

Typeset by Stallion Press
Email: enquiries@stallionpress.com

Four-dimensional paper constructions after Möbius, Klein, and Boy

Eiji Ogasa

Klein

Möbius

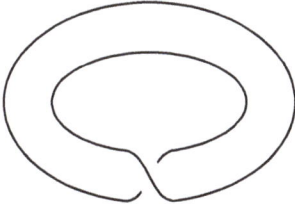

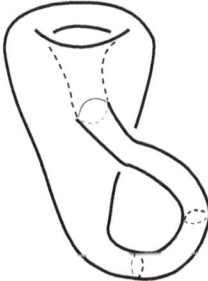

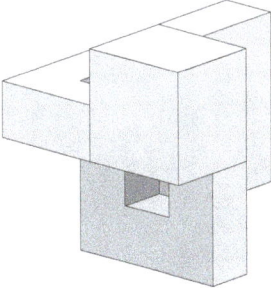

Boy

Girl

Introduction: If You Start on Earth and Travel Straight Up into the Universe, Where Will You End Up?

While reading this book, you will open a door into another dimension via four-dimensional paper constructions. Enjoy it! These paper constructions follow from the work of great mathematicians, such as Möbius, Klein, Boy, Hopf, and others. By doing these paper constructions, you can visualize four-dimensional space and beyond!

If you are interested in mathematics, physics, or just sci-fi novels, movies, and comics, you have likely heard of the Möbius band and the Klein bottle. The more mathematically inclined reader may have also heard of the Boy surface and the Hopf link. While reading this book, you will construct them with your own hands! And after constructing these objects, you will find them even more fascinating.

While they appear in sci-fi stories, the Möbius band, the Klein bottle, Boy surface, and the Hopf link are sophisticated mathematical objects. Throughout the course of this book, the study and creation of these objects will transport you to higher-dimensional spaces. Even more so, this book will help you solidify your foundations in certain areas in mathematics and physics, in particular, topology.

The content of this book is widely accessible. The target audience is first- and second-year undergraduate students in mathematics-related fields, though high school students interested in mathematics and physics should have sufficient knowledge to work through it. If you are familiar with higher-dimensional spaces from a background of loving sci-fi stories, you may find the four-dimensional pictures in this book more intuitive. And while this book is geared toward beginners, its aim is to help the reader build enough topological intuition to be able to tackle more sophisticated areas of mathematics and physics.

Let's begin with a question about the world we live in. Let's say you get in a spaceship right here on Earth and travel straight up into the sky and out of the atmosphere; from the moment you leave the ground, continue traveling straight in the same direction. Where will you end up?

Is there some sort of "boundary" or "end" of the universe? If so, then what is beyond that boundary? Is there something there? If there lies anything tangible beyond the boundary, then clearly it was not a boundary at all; it seems that the "end" was not actually the end. So then what? Does your trip continue on, *ad infinitum*? Is the universe infinite in every direction? Is the size of the universe infinite or finite? It may strike you as odd that there is no physical "end."

Perhaps the answer is as follows: Maybe you travel in one direction and eventually arrive back where you started. Can you imagine what happens during the course of this trip?

Perhaps we could frame it another way, as a sci-fi story. Imagine you have a super-powered telescope, such that you could see immensely far. Let's try to use our super telescope to see as far as possible. What do you see? What does "infinity" look like? See Figure 1 — a recreation of what this might look like in our sci-fi universe.

We can also think of an analogue of this trip in one lower dimension. At some point on Earth, in a wide open field, pick any direction and walk that way. Now just go and go and go. What happens? You get back to the starting point! (If you don't think too hard about the details....)

Looking down at your feet, it appears that the ground is flat and that perhaps the ground is an infinitely wide plane. But we know that the ground is actually a sphere (or rather, very close to a sphere),

Figure 1. What does she see?

that is, the earth. See Figure 2. Thinking about the shape of the earth, then, what shape might the universe be? In this model, the size of the universe is finite. Can you see this model in your mind?

We may consider an analogue of this phenomenon of our space trip in two lower dimensions, namely, traveling around a circle. Can you draw the universe as we have described it in this introduction?

After going through some four-dimensional paper constructions (Chapters 4–7), you will understand it. Later on in this book,

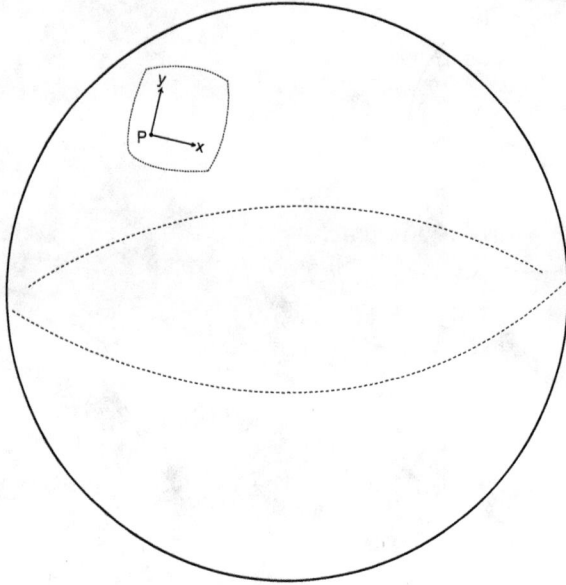

Figure 2.   Earth

we draw a picture of the universe assuming this "looping" condition (Chapter 8).

While reading this book, the reader will become very familiar with four-dimensional space, taking a deep dive into the study of manifolds and topology.

About the Author

The author researches high and low dimensional knot theory, topology. The readers can find his papers by typing in Eiji Ogasa in the search engine. He received PhD of mathematics from University of Tokyo.

He puts movies in YouTube that show paper constructions and DIY constructions exposed in this book. You can find them by typing in Eiji Ogasa in the search engine. The videos must help the beginners understand four and high dimension drawn in this book.

Acknowledgements

The author would like to thank the following people:

Louis H. Kauffman
Liza Jacoby
Alexander Simons
Atsushi Mochizuki
Takayuki Nagasawa
Hiroyuki Yokoi
Saki Kimura
Misaki Watanabe
Kokone Seya
Nanao Seya
Yukako Iwanaga
ittou meiko
Keiichiro Iwai
Misa Iihara
Ayu Yamamoto
Tamaki Imamura
Natsumi Yamda
Nao Mitsuhashi
Sawa Uchida
Hatsune Wanibuchi
Yuika Imai
Yukino Saito
Colin Adams

The names are written in no particular order.

Contents

Prologue: Time Machine and Teleportation

1. A Sci-Fi Detective Novel

Let's begin with a science fiction detective novel.

A science fiction detective novel is an unsolved mystery set in a sci-fi universe. In order to solve the mystery, we can't necessarily use any gadgets we'd like. We have to stick to particular gadgets which were invented in our sci-fi universe: We write the gadgets explicitly in the story.

Don't worry. You will solve this mystery without much difficulty.

Our story is set in a time period far in the future. In this universe, time machines have recently been invented — so recently that only a few people know how to use them.

These time machines are constructed to shrink down small enough to fit in a pocket while not in use. When they are activated, they enlarge to the size of a motorcycle — and jumping on the motorcycle is precisely how to use one of these time machines. In either form, they are remarkably lightweight — light enough for the user to carry around.

2.   A Locked-Room Mystery

Mr. L is a burglar who is the head of an evil organization. His arch nemesis is the detective Mr. H. But Mr. H finally won — one day he caught Mr. L. We detail that fateful day.

Mr. H went out to a wide open field one day and began constructing a big warehouse. When he finished building the structure, Mr. L wandered in to see what it was all about. That's when Mr. H shut the door and locked it from the outside, trapping Mr. L in the warehouse!

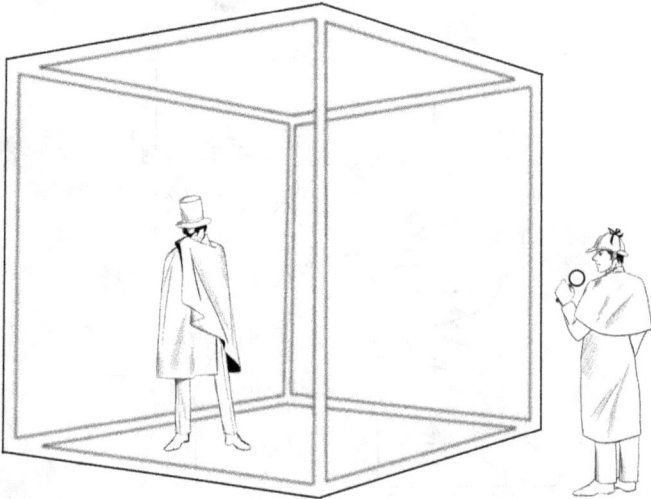

A locked-room

From outside the warehouse, Mr. *H* called out to Mr. *L*, "It has been months since another soul has passed through these fields — and I wouldn't count on that changing any time soon! Now that you're stuck in there, I can see my plan through: I will disguise myself as you and sabotage your organization! I will return once I have successfully taken down your evil group once and for all. Only then will you be released."

Trapped in a warehouse, his organization at risk of collapse — what will be Mr. *L*'s fate?

As luck would have it, Mr. *L* holds one of the newly invented time machines in his pocket.

Using a time machine, one can travel to whatever time they like, either in the past or the future. However, there is a caveat. The time machine takes the user backwards or forwards in time to the same physical location at which they activated the machine. If the user wants to see a particular location at a given time period, they must carry the time machine to that location. Recall that in our sci-fi universe, these time machines are miraculously light enough for the user to carry wherever they may please.

Given all this, what can we say about Mr. *L*'s predicament? Can he escape the warehouse? Can he face Mr. *H* again?

Note that Mr. *H* designed the warehouse with only one door, which he locked from the outside when Mr. *L* wandered in. There are no windows or any other holes in the warehouse structure. What's more, the walls, ceiling, and floor are made of solid iron!

But, prior to that fateful day when Mr. *H* began his construction of the warehouse, there was nothing in the field — and no passersby, either. Mr. *L* knows this since Mr. *H* told Mr. *L* this when Mr. *H* captured Mr. *L*.

3. Challenge to the Reader

If you were Mr. *L*, how would you escape the warehouse?

If you are a fan of science fiction, you may come to the solution quite easily. If you lean into the idea of Mr. *L*'s time machine, then I believe you can solve this problem!

Note: Mr. *L* does not have lots of muscle power nor does he have any weapons to try to break through the warehouse door from the inside. He also does not have supernatural powers, e.g. he cannot phase through the walls like a ghost. The only tools available to him are his time machine and his brain.

The time machine has worked well in the past, so Mr. *L* trusts it and is willing to give it a go to help him out of this conundrum.

You may be concerned that the existence of a time machine implies some paradoxical phenomena in the universe we've imagined. For now, we'll ignore these kinds of concerns. After all, Mr. *L*'s problem is one of science fiction.

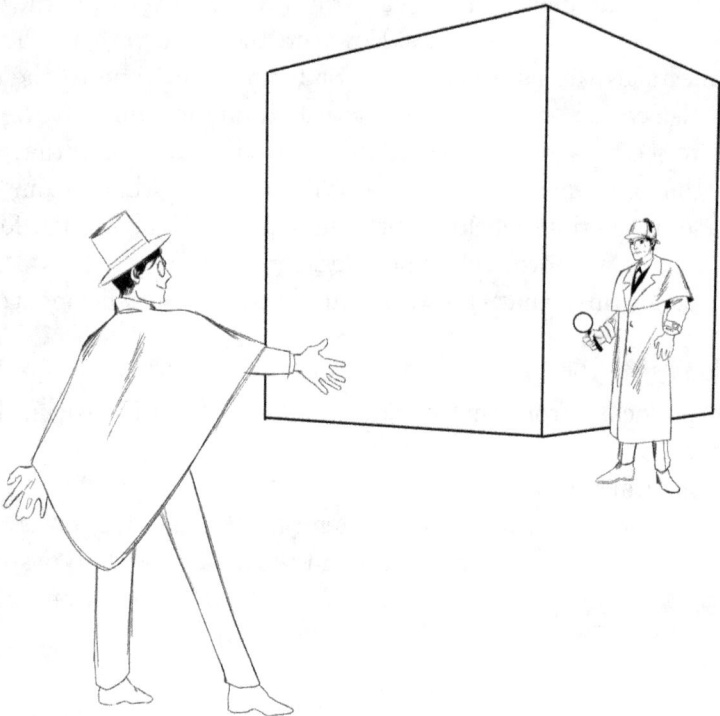

4. Solution

Mr. *L* can escape the warehouse as follows.

First, Mr. *L* travels back in time to one day (or more than one day if he so pleases) before he got caught by Mr. *H* — more specifically, to at least one day before Mr. H built the warehouse. Now, there's nothing and no one in the field; the warehouse has yet to be built and Mr. H is nowhere in sight. Mr. L then walks with his time machine to a location that is outside the boundary of where the warehouse will be in the future. Activating the time machine again, he travels back to the present day — the day he gets caught by Mr. H — appearing as though he has teleported out of the warehouse. All he has left to do is to go find Mr. H, who ought to be astonished at Mr. L's sly escape.

Did you solve it on your own?

You can also consider the following operation. In order to specify a point in the plane, we need two numbers: length and width. In the world we live in, we need length, width, and height to pick out a particular point. Since we need two numbers to determine points on the plane, we say that the plane is *two-dimensional*, and likewise, since three numbers determine points in the space of our daily lives, we say that that space is *three-dimensional*. We explicate the term *dimension* in Chapter 1.

Draw a circle in a plane as in Figure 1.

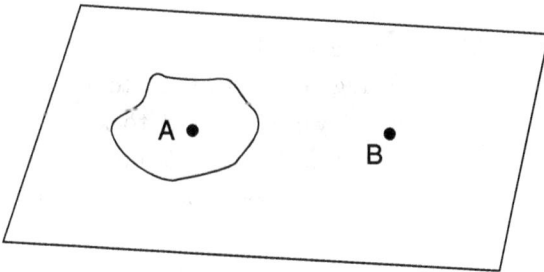

Figure 1.   A circle and two points in the plane

If we smoothly transform the circle without forcing it to cross itself, we still call the resulting figure a circle, even if it doesn't look as we may typically expect. Now, take a point A inscribed in the circle and a point B outside of our circle. Can you move A over to B without crossing over the circle? You may already thinking that we cannot. And you would be right — this is impossible.

However, what if we aren't bound to the plane? What if we can travel above or below it? Then we can easily move point A to point B. Figure 2 shows a path between A and B that goes above the plane containing the two points. While we cannot use such a path in the plane, this is perfectly legal in a space, where a space means a space of our daily lives.

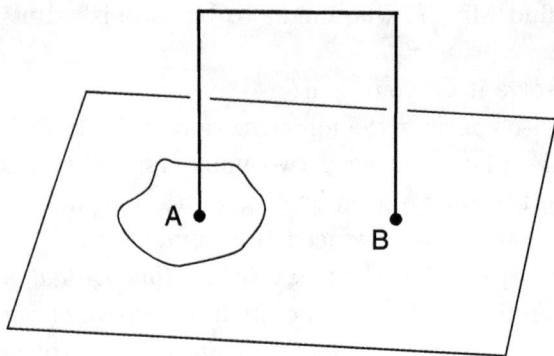

Figure 2.   Using the outside of the plane

Generalize this story and consider an analogue one dimension higher. This would be having a sphere in the space with a point A inside and B outside. Trying to get from A to B is not possible if we stay in the space, but if we're allowed to leave the space, this is suddenly okay. This is precisely what is occurring with Mr. L's escape from the warehouse. The warehouse is the sphere, and Mr. L is trying to travel from a point A inside the sphere to a point B outside the sphere. The time machine enables Mr. L to employ the time axis in the same way that we made use of the height axis when trying to escape the circle in the plane. Put otherwise, we cannot connect a point in the warehouse with a point outside it in three-dimensional space but we can in four-dimensional space. We discuss the concept of four-dimensional space in Chapter 1, following this section. We detail the trick of this sci-fi detective story by using four-dimensional space in Chapter 6.

Chapter 1

Seeing Four-Dimensional Space

1.1. $\mathbb{R}^4$

In this section, the reader will grow familiar with the key concept of four-dimensional space $\mathbb{R}^4$. You may be familiar with four-dimensional space from science fiction in stories like that of Mr. L and his time machine. But four-dimensional space is not only an object in fictional stories, $\mathbb{R}^4$, and even higher-dimesional spaces are serious objects of study for mathematicians and physicists. With this book, you can begin a serious mathematical investigation of high-dimensional spaces.

In the space we live in, we can specify particular points in space by three numbers: the length, the width, and the height. See Figure 1.1.

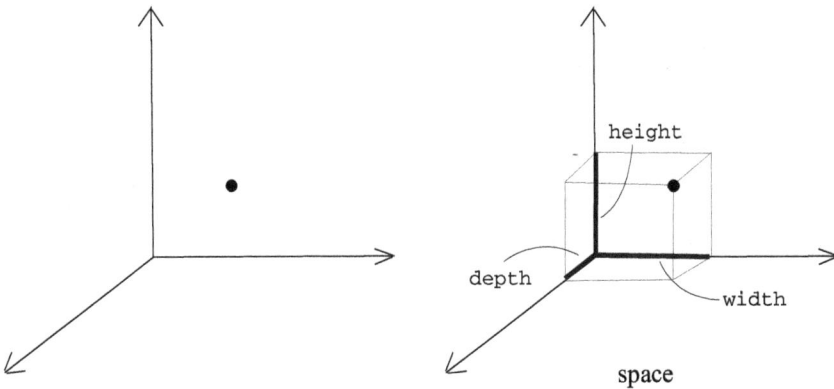

Figure 1.1.

You may be wondering about the time coordinate — whether or not we ought to consider the time at which a point passes through a particular location in space. But don't mind time for now. Let's consider a singular moment in time. And don't worry about whether or not we can really stop time — because we can imagine it! Such fantasization is critical for motivating certain topics in mathematics and physics.

So, in our moment frozen in time, the place of a point in the space we live in is determined by three numbers: width, depth, and height. We can choose these three numbers however we wish, as they are all independent of one another. We thus call this space *three-dimensional space* $\mathbb{R}^3$. Note that the number of dimensions is equivalent to the number of independently chosen numbers which determine the position of a point in space. This is true in general. Thus, we call a line (respectively, a plane) *one* (respectively, *two*) *dimensional space* $\mathbb{R}^1$ (respectively, $\mathbb{R}^2$).

Now, consider not only the width, depth, and height in a singular moment, but consider the time coordinate as well. That is, if lightning flashes at a given place at a particular moment, this instance is specified by four coordinates: width, depth, height, and time.

The space where each point is specified by four numbers is called *four-dimensional space* $\mathbb{R}^4$. In the space we live in, spacially we have width, depth, and height, but we also exist subject to time. When we consider an instance as happening at a location *and* time, we regard the space we live in as four-dimensional space.

We can draw $\mathbb{R}^4$ as in Figure 1.2. Note that there is an axis that is not any of the three in $\mathbb{R}^3$. This axis is oriented differently from the x-, y-, and z-axes. While this axis does not always represent the flow of time, relating it to our world in that way can demystify the fourth coordinate of a point in $\mathbb{R}^4$.

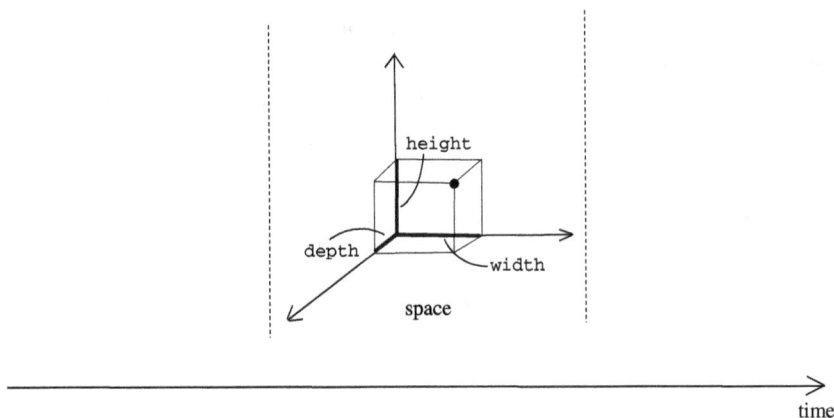

Figure 1.2.

Figure 1.2 is a conceptual illustration of the fourth axis. We draw only one copy of $\mathbb{R}^3$, but really there are infinitely many — as many copies as there are moments of time! You can draw as many as you like, as is exhibited in Figure 1.3, where we illustrate many copies of $\mathbb{R}^3$ to help illuminate our intuition for $\mathbb{R}^4$.

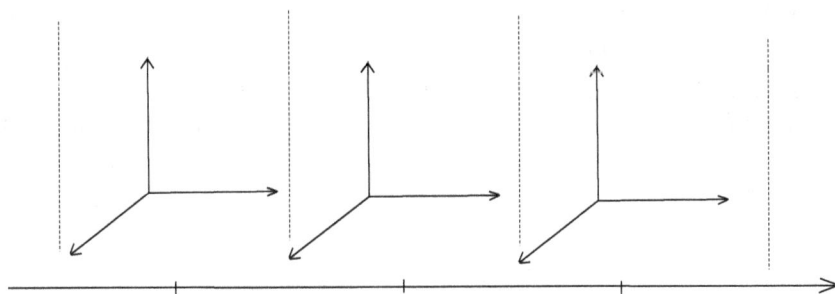

Figure 1.3.

We continue to regard the fourth axis as time, but in many cases, another metric — such as temperature or humidity — may be more appropriate to specify the position of a point.

But for now, we stick to thinking about the fourth coordinate as a specified moment in time. In general, though it is okay to use any unit of time, you may assume that the unit is seconds. In Figure 1.4, we draw $\mathbb{R}^3$ at times $t = 1, 2, 3, 4, 5$.

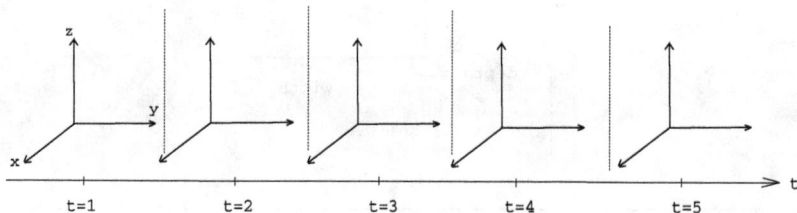

Figure 1.4. $\mathbb{R}^4$

You have likely noted that four-dimensional space is regarded as the locus taken by three-dimensional space when moving in a direction that is independent of width, depth, and height. We can imagine this movement in our heads.

Width, depth, height, and time are generally the four numbers we use to specify the place of a point in $\mathbb{R}^4$. However, as aforementioned, we don't need to discuss the fourth axis as time. In fact, mathematically, we may consider each of the coordinates as mere four numbers.

Consider $\mathbb{R}^4$ where the position of a point in $\mathbb{R}^4$ is determined by (x, y, z, t): width, depth, height, and time. Suppose a light is on from $t = 2$ to $t = 4$, as shown in Figure 1.5. The trajectory of the light is an object A. What is A in $\mathbb{R}^4$, as shown in Figure 1.5?

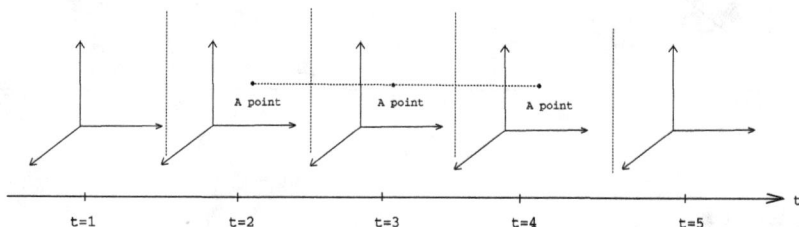

Figure 1.5.   What is A?

You may recognize A as a segment. Can you see this as a figure living in four-dimensional space?

Furthermore, we can move A around in $\mathbb{R}^4$. The result is shown in Figure 1.6. Can you see this movement in $\mathbb{R}^4$?

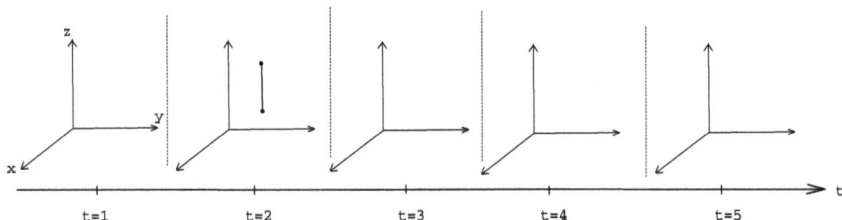

Figure 1.6. *A* is ...

In the following, we describe the process of moving a segment around in $\mathbb{R}^3$. Moving around segments in $\mathbb{R}^4$ — like moving the one shown in Figure 1.5 to what's shown in Figure 1.6 — is an analogue of the same process, but one dimension higher.

Draw a segment in $\mathbb{R}^3$, where the position is specified by three numbers, x, y and t, as shown in Figure 1.7. Consider the planes perpendicular to the t-axis intersecting the segment, shown in Figure 1.8, and the sections of our segment contained in these intersecting planes. See page 15 for the term, *intersect*.

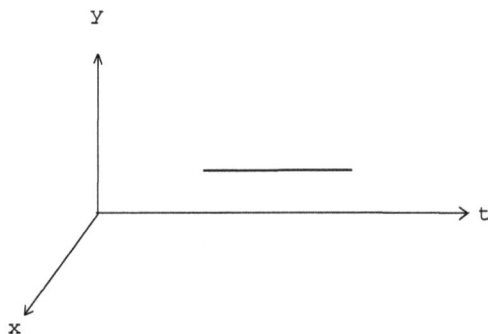

Figure 1.7. A segment in $\mathbb{R}^3$

Sections $\mathbb{R}^3$ at $t = 1, 2, 3, 4, 5$ in $\mathbb{R}^4$ of Figure 1.5 correspond to sections $\mathbb{R}^2$ at $t = 1, 2, 3, 4, 5$ in $\mathbb{R}^3$ of Figure 1.8.

Figure 1.5 is a one-dimensional higher analogue of Figure 1.8.

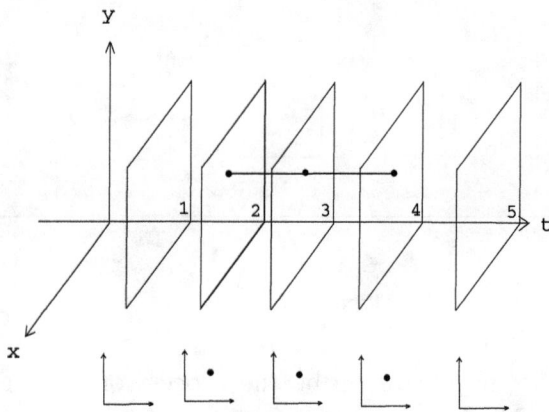

Figure 1.8.   Sections of a segment in $\mathbb{R}^3$

A movement from Figure 1.5 to Figure 1.6 is a one-dimensional higher analogue of that from Figure 1.9 to Figure 1.10.

We move the segment from Figure 1.9

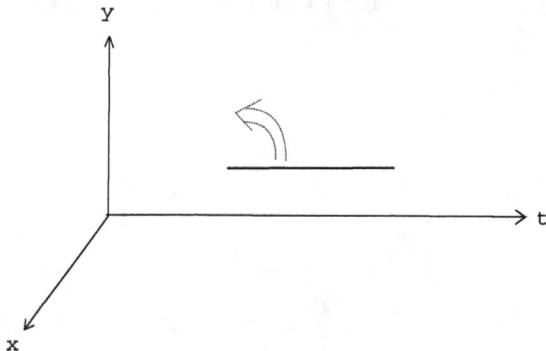

Figure 1.9.   Moving a segment in $\mathbb{R}^3$

and place it as it appears in Figure 1.10.

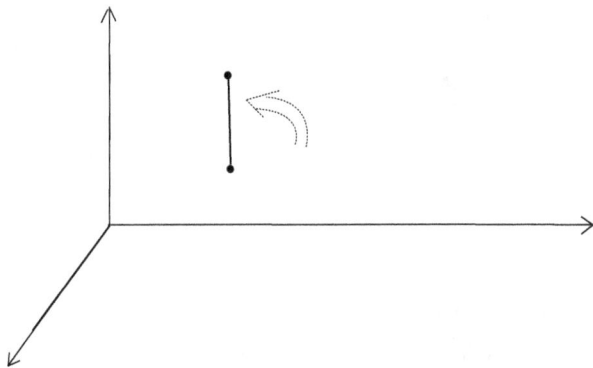

Figure 1.10. The result of the moving

Since the physical space we live in is three-dimensional, mathematics is much more intuitive in four or greater dimensions than in three or fewer. Consequently, when we fantasize about objects in high-dimensional space, a lower-dimensional analogue can help ground our fantasies.

1.2. Getting Out in Four-Dimensional Space

We explain the escape of Mr. L in §4. In Figures 1.11–1.17, we draw the route of his escape in $\mathbb{R}^4$, where the position of every point along the path is specified by width, depth, height, and time.

Recall that the 'four-dimensional escape' we just outlined is merely a one-dimension-higher analogue of the process described associated with Figure 2 on page xxii, which is perhaps more visualizable.

In Figure 1.11, we regard Mr. L as a point and the warehouse as the sphere. From the problem statement, we gather that the sphere surrounds the point and the whole system lives in $\mathbb{R}^3$ at time coordinate $t = $ today.

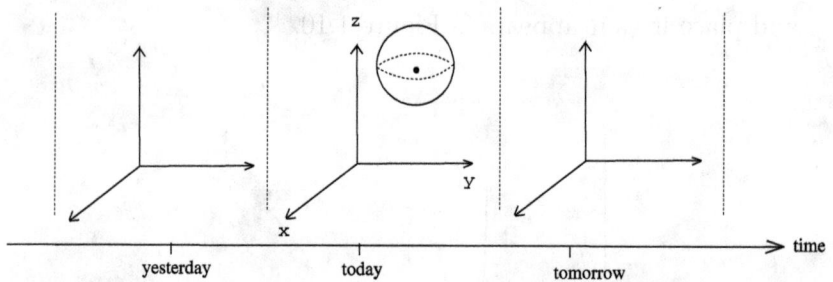

Figure 1.11.

Of course, there are four-dimensional space $\mathbb{R}^4$ and the time axis before yesterday and after tomorrow although we draw only a part of $\mathbb{R}^4$. We need such an imagination.

Mr. L moves from $\mathbb{R}^3$ at $t = $ today to $\mathbb{R}^3$ at $t = $ yesterday. He has only moved along the time axis — the width, depth, and height coordinates remain fixed. This movement is shown in Figure 1.12.

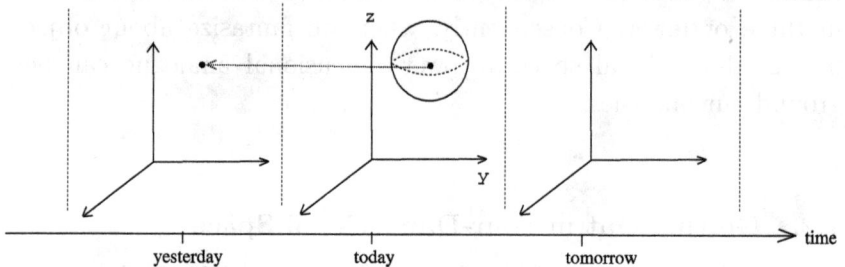

Figure 1.12.

Of course, there are spaces after tomorrow and before yesterday although they are not drawn. Things that we are interested in now exist only in (yesterday) $\leq t \leq$ (today). Imagine!

In Figure 1.13, we see the endpoint of the movement shown in Figure 1.12. Mr. L now stands at this endpoint.

In order to escape the warehouse, Mr. L must move from the point in $\mathbb{R}^3$ at $t = $ yesterday shown in Figure 1.13 to some area outside of

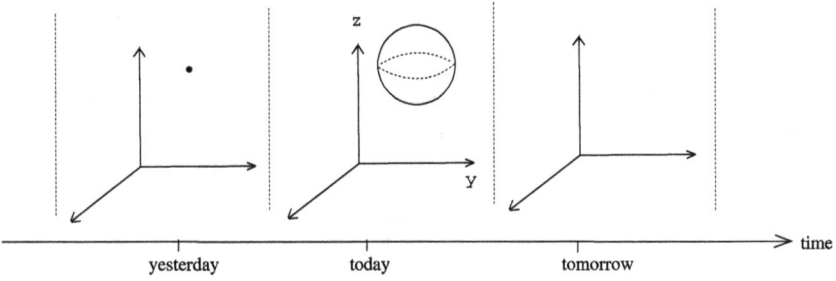

Figure 1.13.

where the warehouse is at time $t = $ today. We show this movement in Figure 1.14.

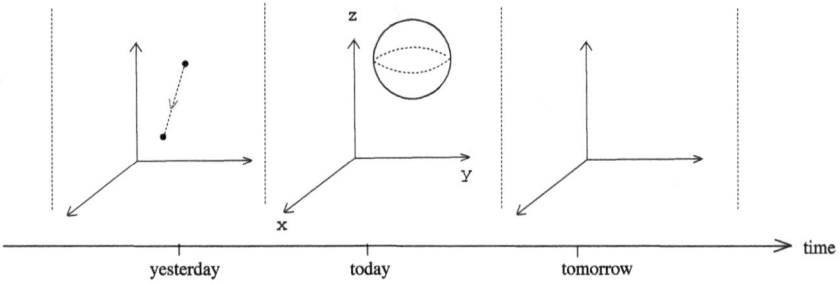

Figure 1.14.

In Figure 1.15, we show the endpoint of the movement shown in Figure 1.14. Mr. L now stands here.

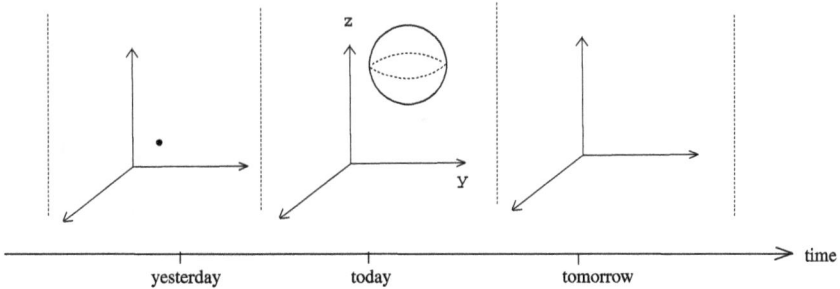

Figure 1.15.

Now, in order to escape the warehouse, Mr. L can move from the point at $t =$ yesterday shown in Figure 1.15 to the time coordinate $t =$ today while keeping the width, depth, and height coordinates fixed. We show this trajectory in Figure 1.16.

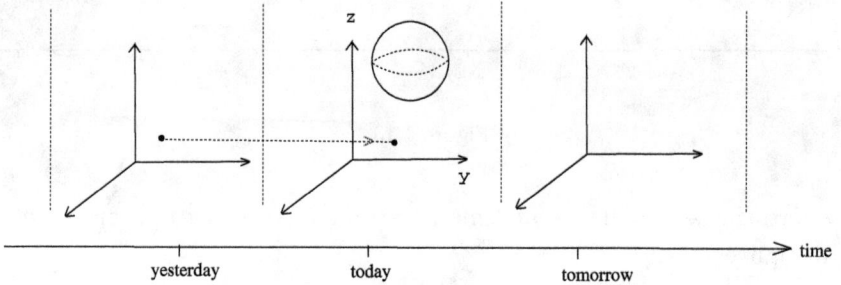

Figure 1.16.

In Figure 1.17, we include the endpoint of the movement shown in Figure 1.16. By returning to this point, Mr. L finds himself outside the warehouse in the present day.

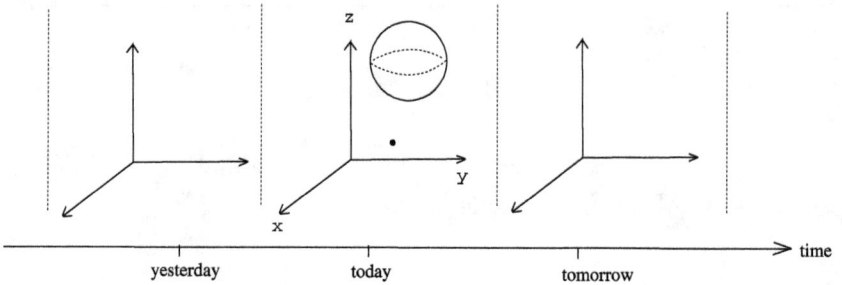

Figure 1.17.

In Figure 1.18, we show the full trajectory of Mr. *L*'s escape.

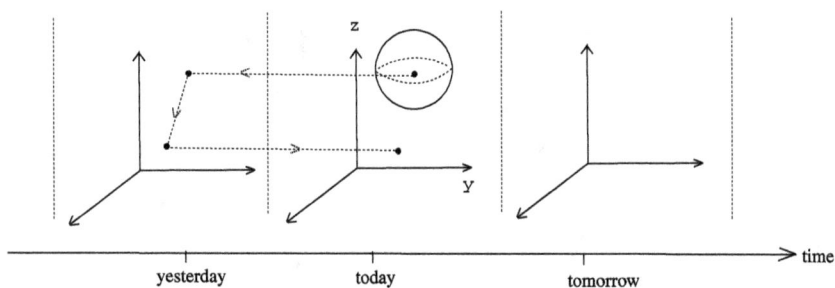

Figure 1.18.

Chapter 2

Möbius Band

2.1. Let's Construct Möbius Band

In this section, we introduce a new type of object after explicating a more intuitive one.

Let's draw a horizontal rectangle as drawn in Figure 2.1. Note that this rectangle includes its bounding edges and the body.

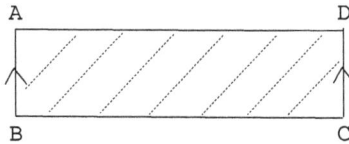

Figure 2.1.

Glue the segments AB and DC as drawn in Figure 2.2 so that the arrows coincide. Then A and D meet and B and C meet.

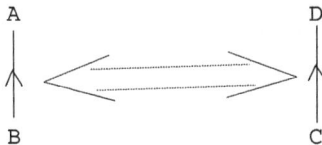

Figure 2.2.

Except for segments AB and DC, the rectangle does not touch itself. The resulting object does not touch itself. What figure

13

do we obtain? In what follows, we define what we mean when we say an object does or does not "touch itself."

Given an object X, if a point P in X touches a point Q in X and $P \neq Q$, we say that X *touches itself*. This is shown in Figure 2.3.

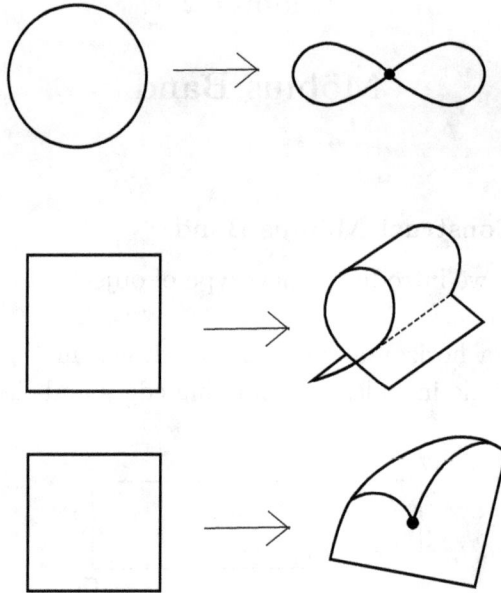

Figure 2.3. First row: A circle without the body touches itself at one point. Second row: A square including the body touches itself at one segment.
Third row: A square including the body touches itself at one point

Assume that we are given a figure X. Determine that we identify which points in X with which points in X, where there may be infinitely many such points and the points may make a figure, e.g. a segment, a circle, etc. We often consider what new figure Y these identification construct from X.

Suppose we constructed Y in a space V, e.g. 3-dimensional space. Assume that identified points in Y are only the points that we determined for X beforehand. Then we say that the new figure Y made from X does not touch itself. One way of saying, Y is determined *abstractly* when we say which points are identified with which points.

In the case of the above example, the rectangle touches itself. The resulting figure that we are making will not touch itself.

In the latter pages of this book, we consider whether we can construct Y without touching itself in V, in more difficult situations.

Furthermore, we introduce a term, *intersect*.

We say that in the upper figure of Figure 2.4, curves AB and CD touch but do not intersect.

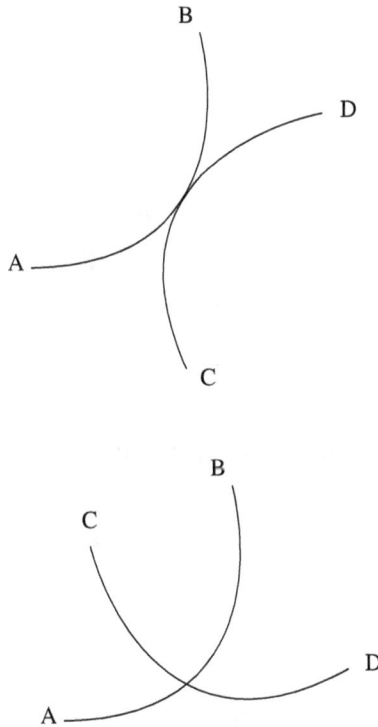

Figure 2.4. Intersecting and touching

We say that in the lower figure of Figure 2.4, curves AB and CD touch and intersect.

See a point that is included in both curves. If a curve goes over the opposite side of the other curve, then we say that the two curves *intersect*.

Here, we define in the case of two one-dimensional objects. In other cases of two different dimensional objects, we can define in the same fashion.

Note that the words 'touch' and 'intersect' are defined rigorously in mathematical literature.

We perform the construction shown in Figure 2.5.

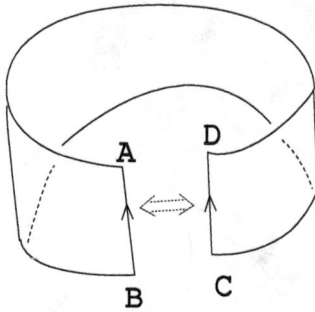

Figure 2.5.

We obtain the object shown in Figure 2.6, called an *annulus*, or *cylinder*.

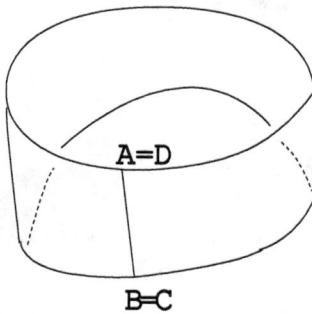

Figure 2.6.

As shown in Figure 2.7, we can stretch and/or bend the object continuously without cutting or touching itself, as shown in Figure 2.6. The object we obtain is still an annulus.

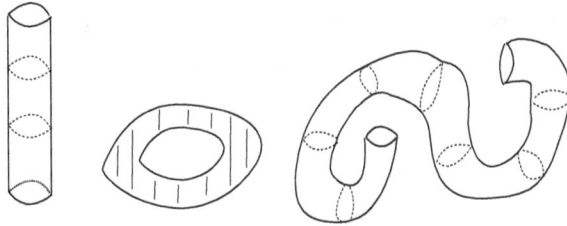

Figure 2.7.

We now construct this section's main object of interest. We once again draw a horizontal rectangle containing both the boundary and the body, as shown in Figure 2.8.

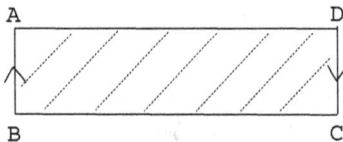

Figure 2.8.

Attach the segments AB and DC as drawn in Figure 2.9 so that the arrows coincide. Note that the arrow on the DC edge in this case is oriented oppositely from that in the case of the annulus (Figure 2.2).

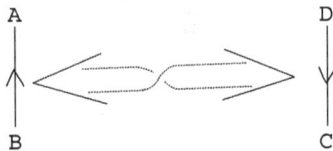

Figure 2.9.

We see that A and C meet and B and D meet. Except for the AB and DC segments, the rectangle does not touch itself. The resulting object does not touch itself. What figure do we obtain?

Now perform the construction shown in Figure 2.10. Note that when we glue up the rectangle, we are introducing a "half twist," whereas when constructing the annulus (as shown in Figure 2.5) we did not perform any sort of twisting.

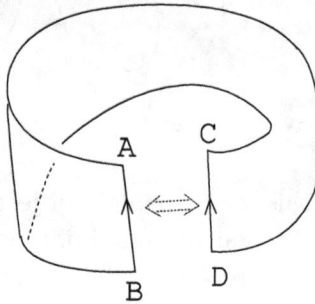

Figure 2.10.

We obtain the object shown in Figure 2.11. This "half-twisted" glued-up rectangle is called a *Möbius band.*

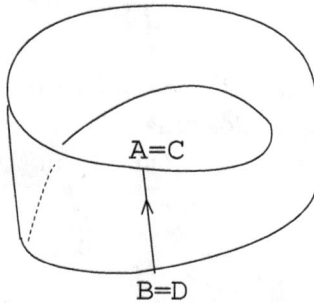

Figure 2.11.

Note the difference between an annulus and a Möbius band. While the "half-twist" is the only aesthetic difference, we explain later on the several different properties that distinguish the two objects mathematically (§6.1–§6.4).

While an annulus is named after a common noun, the Möbius band gets its name from the great mathematician Möbius (1790–1868). Möbius devoted countless hours toward understanding the object to subsequently elucidate its immense allurement and mathematical importance. We honor his work by letting the object to assume his name.

However, it is important to note that Möbius cannot have been the first person in history to construct a Möbius band. Surely, long before Möbius, humans doing paper constructions came across the object by mistake, perhaps trying to construct an annulus and accidentally introducing a half-twist before gluing up the ends.

To make your own Möbius band at home, paper towel may work best — printer paper is a little too thick and tissue paper is a little too thin.

We make more complicated paper constructions later on, but for now we focus on the Möbius band. In this part, we introduce some of its properties, as we eventually use the Möbius band to discuss four-dimensional space.

2.2. Create a Möbius Band with Paper and Scissors

You will need a piece of paper (perhaps paper towel) and a pair of scissors. Create a rectangle from the piece of paper and draw a line down the center of the rectangle (Figure 2.12). Use paper that ink soaks up from front to back.

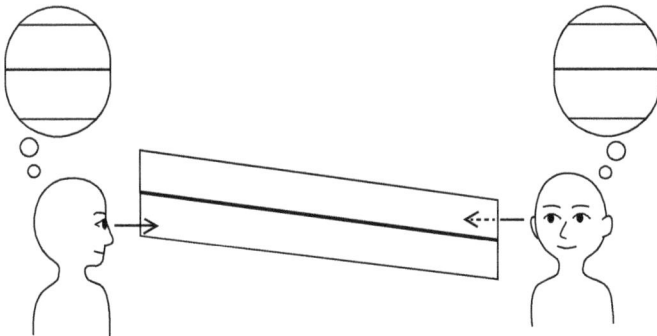

Figure 2.12. Paper that ink soaks up from front to back

Create another rectangle in the same way. Out of the first such rectangle with a line down the center, tape the ends to create an annulus. From the second rectangle, tape the ends to construct a Möbius band. Note that both objects have their center lines as those shown in Figure 2.13.

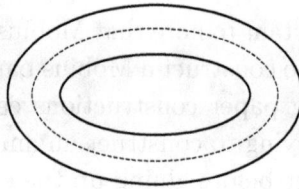

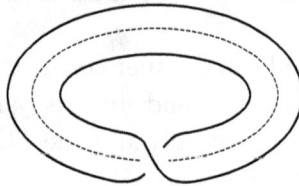

Figure 2.13.

Using the scissors, cut the annulus along the center line, as shown in Figure 2.14. What do we obtain?

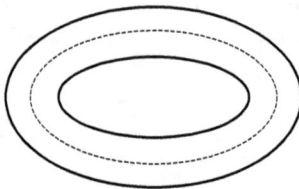

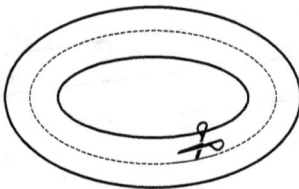

Figure 2.14.

We see that our annulus, when cut down the middle, has become two annuli. This is shown in Figure 2.15.

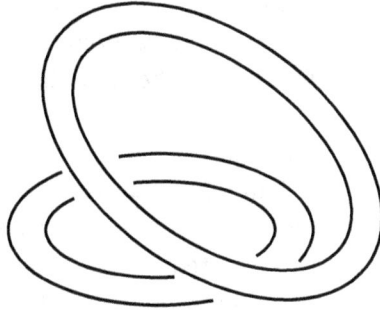

Figure 2.15.

Now cut the Möbius band along the center line as shown in Figure 2.16. What do we obtain?

If you do not know the answer, try this paper construction as drawn in Figure 2.16.

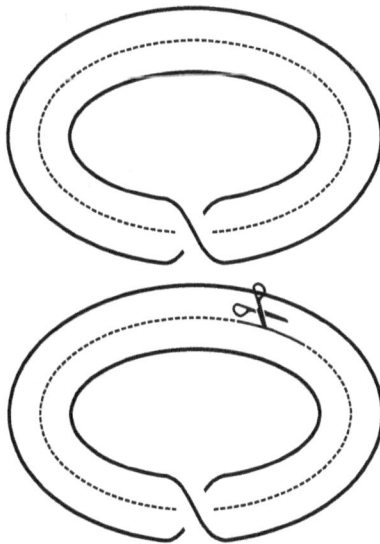

Figure 2.16.

As note on page 19, it is best to use paper towel for this paper construction. Printer paper may be too cumbersome and tissue paper may be too flimsy.

We obtain the object shown in Figure 2.17. Note that this object is one piece — not two! — and that its boundary is a disjoint union of two circles. (In §6.4, we discuss the boundary of an annulus and that of an Möbius band.)

When we construct an object from a long narrow rectangle as done in Figure 2.6, and Figure 2.11, we call the object an annulus or a Möbius band, depending on how many times it's twisted: for an integer n, if the band is twisted n times, it's an annulus; if it's twisted n and a half times, it's a Möbius band.

So, the object shown in Figure 2.17 (which we got from cutting a half-twisted Möbius band down the center line) is an annulus.

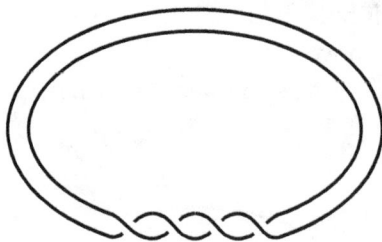

Figure 2.17.

Now take an annulus which is twisted one time, as shown in Figure 2.18.

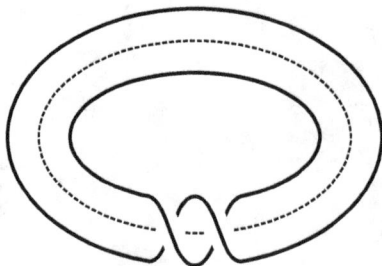

Figure 2.18.

Cut it along the center line as in Figure 2.19. What is the result?

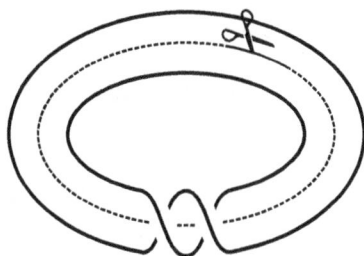

Figure 2.19.

We obtain the object shown in Figure 2.20. It is again a disjoint union of two annuli — yet this time, they are linked, whereas those shown in Figure 2.15, are not.

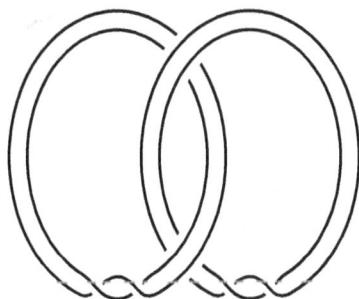

Figure 2.20.

Now take a Möbius band twisted one and a half times, as shown in Figure 2.21.

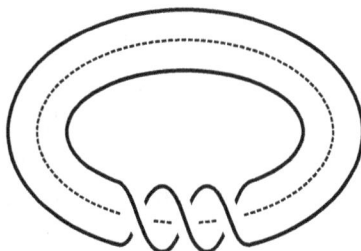

Figure 2.21.

Cut it along the center line as in Figure 2.22. What do we obtain?

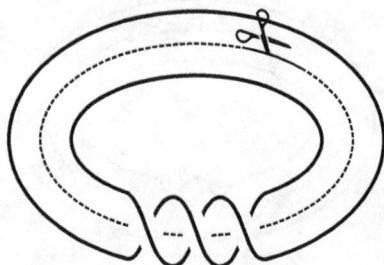

Figure 2.22.

The result is an annulus as in Figure 2.23. Furthermore, the annulus is knotted. Try this paper construction on your own.

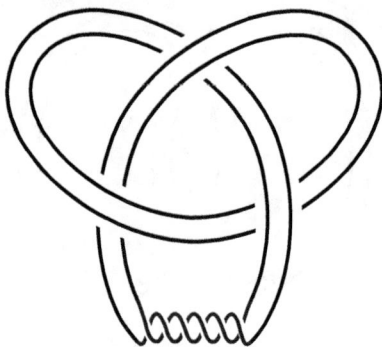

Figure 2.23.

Miscellaneous: Knot Theory

If we have m circles embedded disjointly in three-dimensional space, the disjoint union of the circles is called an *m-component link*. If $m = 1$, the object is called a *knot*. (Note that the words "embed" and "disjoint" are technical mathematical terms.)

You may be wondering: What kind of knots exist? What properties do knots have? These questions are very important to the theory. Mathematicians and scientists in other fields have been investigating these questions for a long time.

We have come across some examples of knots and links in our paper constructions. The center lines of the two unlinked annuli in Figure 2.15, make a link which we refer to as the *trivial two-component link*. More simply, the center line of one annulus is a circle — in this context, we call that in the upper object of Figure 2.13, the *trivial knot*. The trivial knot appears in many other figures in §8. The center line in the lower object of Figure 2.13 is also the trivial knot.

A famous link called *the Hopf link* appears as the center lines of the two linked annuli in Figure 2.20. We introduce the Hopf link in Part 4. In Figure 2.23, we see yet another famous knot. The center line of the knotted annulus is referred to in knot theory as the *trefoil knot*.

Let n be a non-negative integer. Given a long rectangle like the one from our paper constructions, twist it n and $n + \frac{1}{2}$ times before gluing it back up to get either an annulus and a Möbius band, respectively. Cutting the object along its center line yields a link we call the $(2, 2n)$-*torus link*, which is a 2-component link, and the $(2, 2n + 1)$-*torus knot*, respectively.

The object shown in Figure 2.24 is a knot called the *figure-eight knot*. The figure-eight knot cannot be obtained from the above method, telling us that we cannot find all knots just by cutting annuli and Möbius bands.

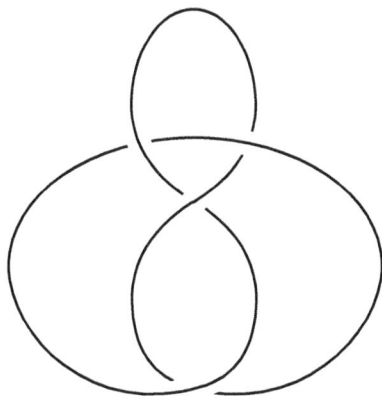

Figure 2.24. The figure-eight knot

Chapter 3

The Torus

3.1. Squares and Tori

We construct a new figure by the following process. Begin with a square $ABCD$ including the boundary and the body. Draw arrows on the edges like those shown in Figure 3.1.

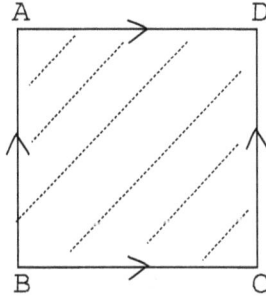

Figure 3.1. A square

Glue together the segments AB and DC so that the arrows coincide as in Figure 3.2. The points A and D meet, and the points B and C meet. Mathematically, we refer to points (respectively, segments) "meeting" as points (respectively, segments) *being identified*.

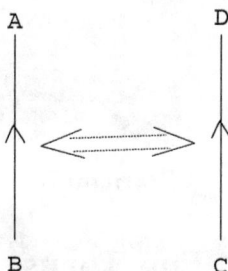

Figure 3.2.

Now attach the segments BC and AD as drawn in Figure 3.3 so that the two arrows coincide. The points B and A become identified with one another, as do C and D with one another.

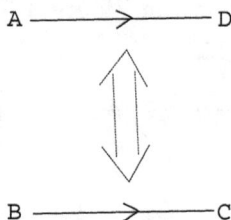

Figure 3.3.

Note that from both of our edge gluings, we have identified all four points with each other. Consequently, the points A, B, C, and D are only one point in our new object.

The square $ABCD$ may also be bent and stretched continuously without cutting or touching itself. Imagine this!

Suppose that the square does not touch itself except for the boundary segments that we have identified together. We suppose that the resulting object does not touch itself. See page 13 for what we mean when we say that the square does not "touch itself."

After making the aforementioned identifications, what do we obtain? It is not so difficult — try it!

First, do as is drawn in Figure 3.4.

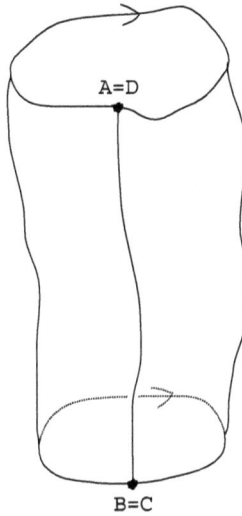

Figure 3.4.

Next, glue the segments as shown in Figure 3.5. The resulting figure looks like a doughnut.

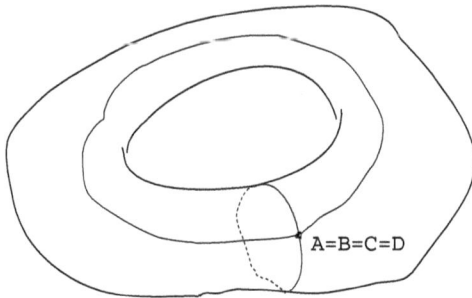

Figure 3.5.

Note, however, that it is only the boundary of a doughnut — it does not contain the body. Rather, it is like an inner tube that is seen in a swimming pool and that does not have the body. We call this object a *torus*.

Chapter 4

The Hopf Link

In this part, we use a paper construction to help familiarize the reader with four-dimensional space. In order to see objects in $\mathbb{R}^4$, you only need a little bit of imagination!

4.1. Two Circles and a Disc

The two linked circles shown in Figure 4.1 make the *Hopf link*.

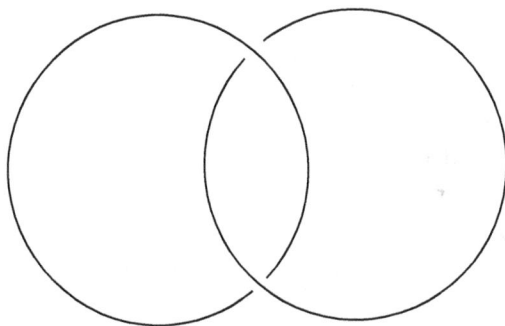

Figure 4.1. The Hopf link

You can construct the Hopf link easily using your fingers! This is shown in Figure 4.2.

Figure 4.2.   The Hopf link made by fingers

Hopf is the name of a great mathematician who made exciting strides researching the link in Figure 4.1, so we have named the link after him. However, Hopf cannot have been the first person to construct the Hopf link, especially given how easy it is to construct with your fingers! It must be the case that people long before Hopf had constructed the link, even if just using their hands.

To one of the components of the Hopf link, attach a disk. We ask the following question.

Question 4.1. Can we attach the disk so that it does not touch the other component of the Hopf link?

Do you think that it is impossible? You may be thinking that the disk and the other component always touch, as is shown in Figure 4.3.

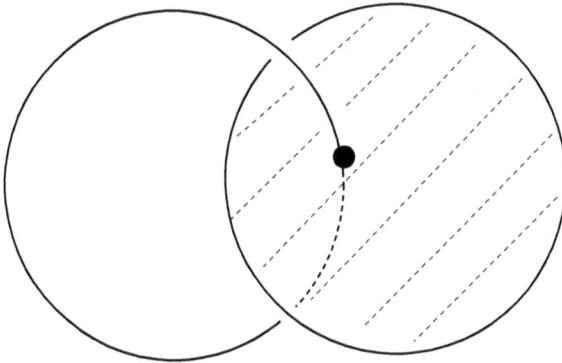

Figure 4.3.

If we curve the disk continuously without cutting or touching itself as in Figure 4.4,

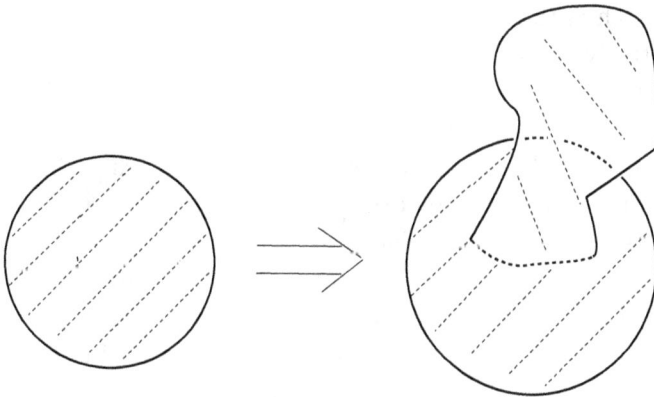

Figure 4.4.

or if we bend it continuously without cutting or touching itself as the left object in Figure 4.5, it seems to be impossible.

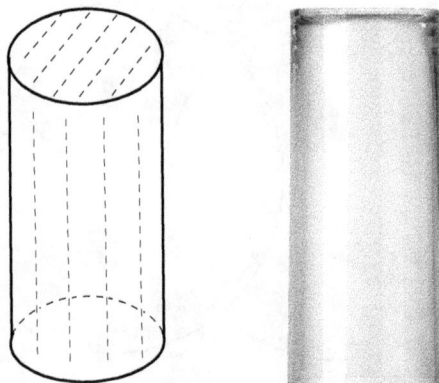

Figure 4.5.   An up-side-down tumbler, made of the side and the top. Note there is no bottom

Note that the left object shown in Figure 4.5 is like a glass or a tumbler, whose side is perpendicular to the bottom, placed upside down, which is the right object put in Figure 4.5.

Intuitively, it seems impossible to attach a disk to one of the circles without touching the other. In this case, our intuition proves to be correct — it has in fact been mathematically proven that in *three-dimensional space* it is not possible to attach a disk into one circle without touching the other.

However, in *four-dimensional space*, we actually can attach a disk to one of the components of the Hopf link without touching the other component. How do you do that?

We show the picture in four-dimensional space in the following sections and answer the above question.

4.2.   You Can Do It in Four-Dimensional Space

We answer the question we posed about the Hopf link (see Figure 4.6) after we exhibit some important figures in $\mathbb{R}^4$ that are useful to answer this question.

We introduced four-dimensional space $\mathbb{R}^4$ in §1.1. Now let's imagine objects in $\mathbb{R}^4$.

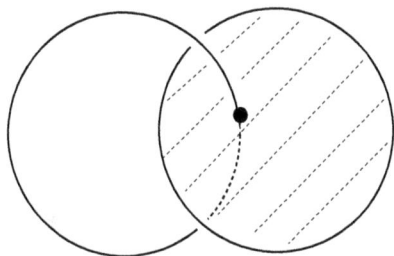

In $\mathbb{R}^4$ we can attach a disc to one circle in the Hopf link so that the disc does not touch the other circle, although we cannot do it in $\mathbb{R}^3$.

How do you do that?

Figure 4.6.   Question 4.1 on page 32. See also page 34

We draw four-dimensional space $\mathbb{R}^4$ and put and move objects in $\mathbb{R}^4$. In the following section, we explain points to be careful of when we draw and see $\mathbb{R}^4$.

4.3.   How to Draw Four Dimensional Space

We begin a preparation of answering the question posed in Figure 4.6, and explain points to be careful of when we draw and see $\mathbb{R}^4$.

Suppose that a ring-shaped light shines at time $t = 1$ second and that it shines for exactly two seconds so that at time $t = 3$ seconds the light turns off.

How to draw this situation? One way of drawing is Figure 4.7. Here, we draw two $\mathbb{R}^3$ at $t = 0$ and $t = 4$ and represent the fact that there is nothing in $t < 1$ and $3 < t$ as in Figure 4.7. However, if we draw $\mathbb{R}^4$ like Figure 4.7, the height is very small and we cannot draw the figure in $\mathbb{R}^4$ so large as we hope.

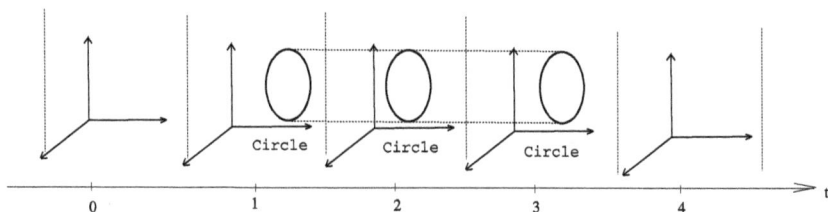

Figure 4.7.

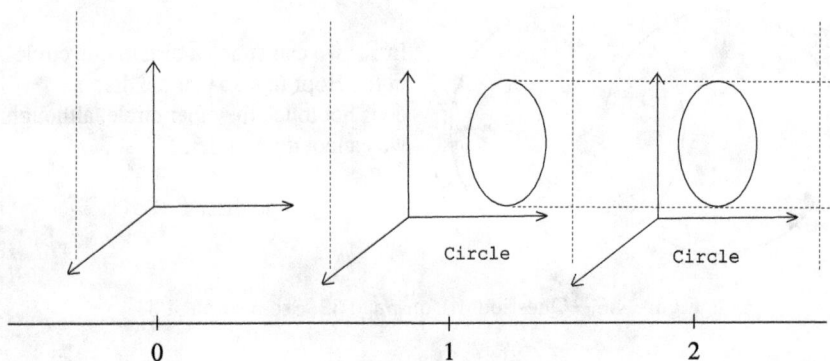

Figure 4.8.   Figures 4.8 and 4.9 make one object

So we sometimes draw it on two facing pages as in a combination of Figures 4.8 and 4.9 in this book. Or, we omit both ends of Figure 4.7 and draw it as in Figure 4.10 so that the height of the figure is larger than that of Figure 4.7. We adopt this kind of omission when such a situation is clear from the context.

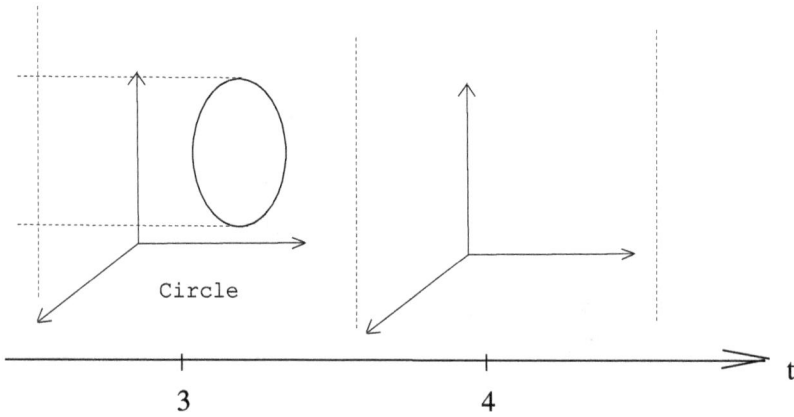

Figure 4.9. Figures 4.8 and 4.9 make one object

Now, consider the trace of the light shining in $\mathbb{R}^4$. What is this object?

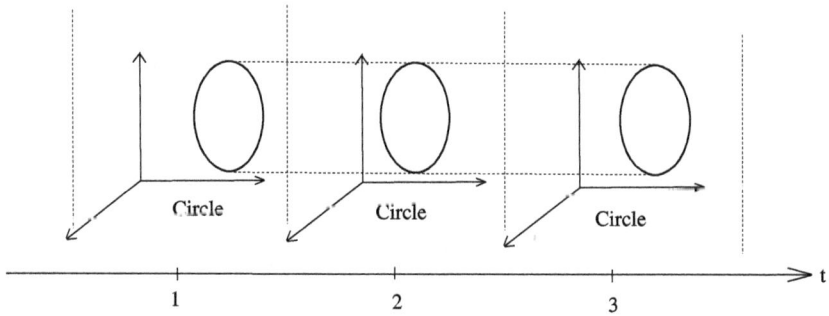

Figure 4.10.

You may understand the object to be like the cardboard around which plastic wrap is wound, as shown in Figure 4.11. Mathematically, we call this the *side* of the *cylinder*.

Figure 4.11.   The cylinder around which plastic wrap is wound

4.4.   Put and Move Objects in Four-Dimensional Space

We continue a preparation of answering the question posed in Figure 4.6.

You may be able to visualize moving the side of the cylinder in Figure 4.10 to the position of that shown in Figure 4.12. Note that this movement is carried out in four-dimensional space. Can you see this movement?

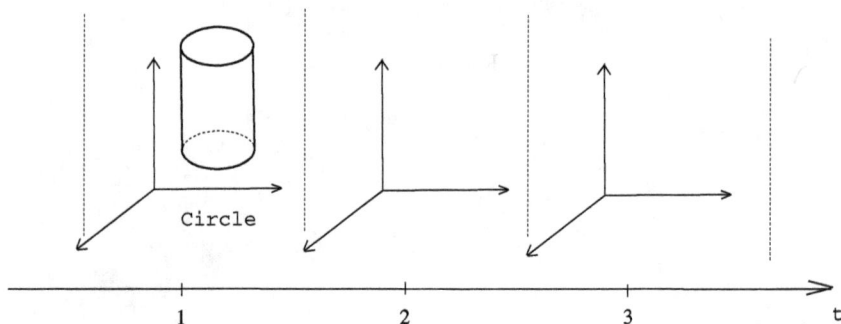

Figure 4.12.

An analogue of this movement one dimension lower is as follows.

Place a cylinder in $\mathbb{R}^3$ to be positioned vertically and horizontally as that in Figure 4.13. Slice the cylinder perpendicular to the horizontal axis; consider the sections of the cylinder we make from this cutting process. Moreover, consider this happening in real time (regard the horizontal axis as the time axis). At first, we have nothing. Then, a circle (just the boundary) appears and remains there for some amount of time. Eventually, the circle ceases to exist. This process forms our cylinder in $\mathbb{R}^3$.

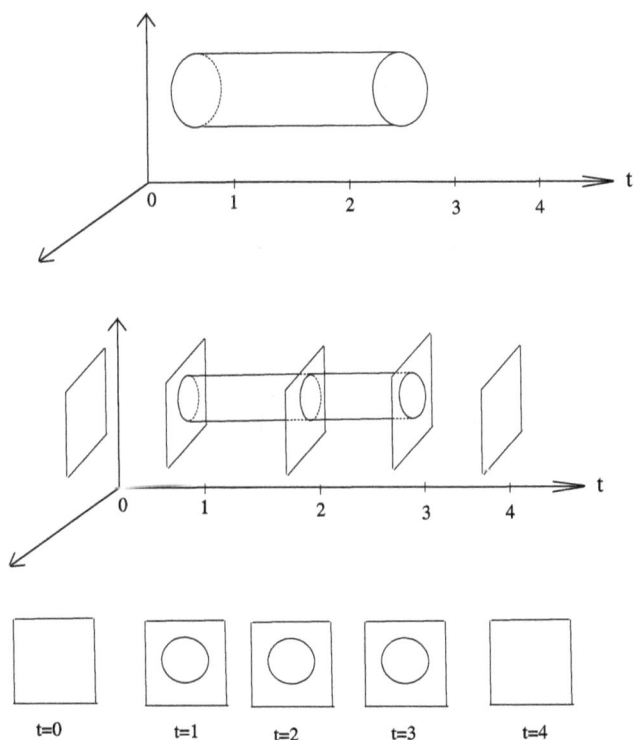

Figure 4.13.

As aforementioned, it is often useful to construct a one-dimension-lower analogue of our four-dimensional motions. We continue on with our three-dimensional analogue, then.

Now, position the cylinder as shown in Figure 4.14.

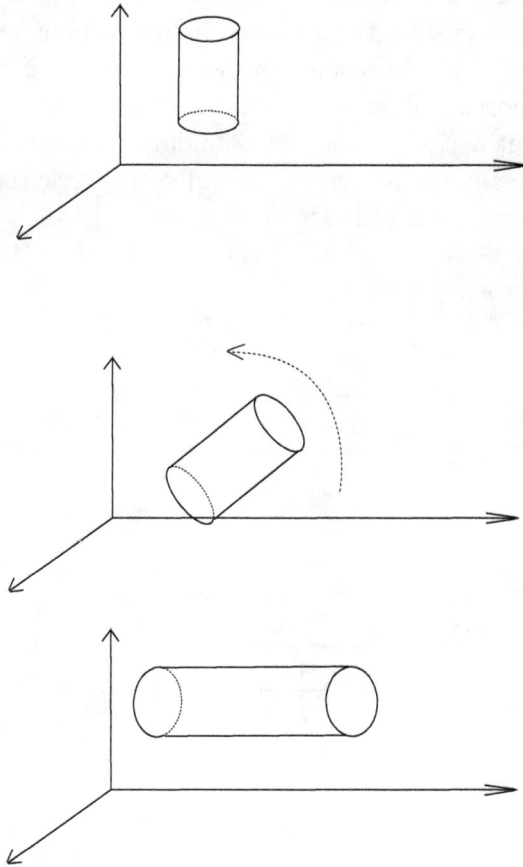

Figure 4.14.

Do you see the analogy?

Let's go through another example. Suppose that a disk-shaped light shines at exactly time $t = 1$ second, remains in the position for two seconds, and then disappears at exactly time $t = 3$ seconds. This process looks like what is illustrated in Figure 4.15. In that case, consider the object traced by the light shining in four-dimensional space. What does this object look like?

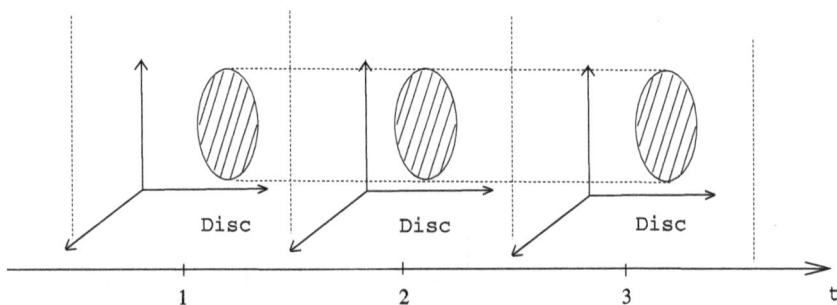

Figure 4.15.

You may have hypothesized that the object is a filled-in cylinder — in which case you would be correct!

Can you visualize transforming the object shown in Figure 4.15 to appear like that shown in Figure 4.16, all in $\mathbb{R}^4$?

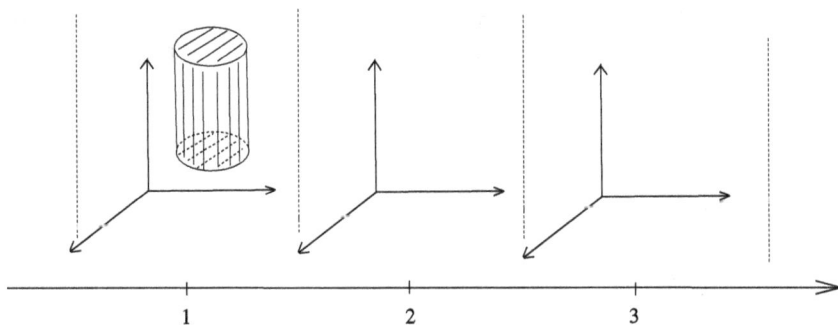

Figure 4.16.

Let's look at another example — similar to the previous two. It is always good to think of analogies and examples to help deepen your understanding.

Suppose a disk-shaped light shines instantaneously at time $t = 1$ second. Moreover, suppose that the boundary circle of the disk-shaped light remains for a duration of two seconds. At $t = 3$ seconds, the body of the disk flashes instantaneously once more.

Figure 4.17 gives you an idea of what this looks like. Can you imagine what object this process traces in $\mathbb{R}^4$?

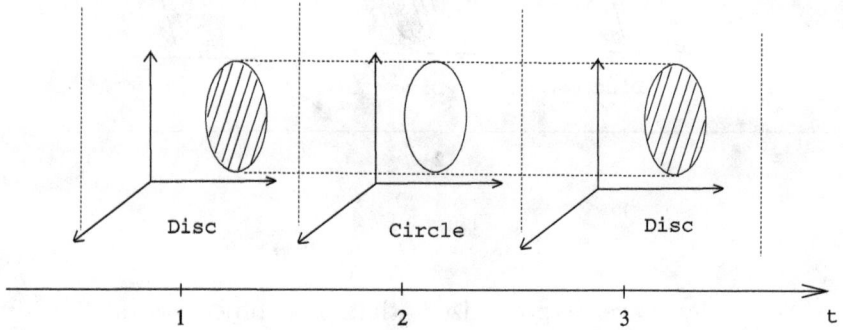

Figure 4.17.

In $\mathbb{R}^4$, we get an empty cylinder — or rather, the *boundary* of a cylinder. Perhaps you can imagine this process in $\mathbb{R}^4$ as the object in Figure 4.17 being translated to the place shown in Figure 4.18.

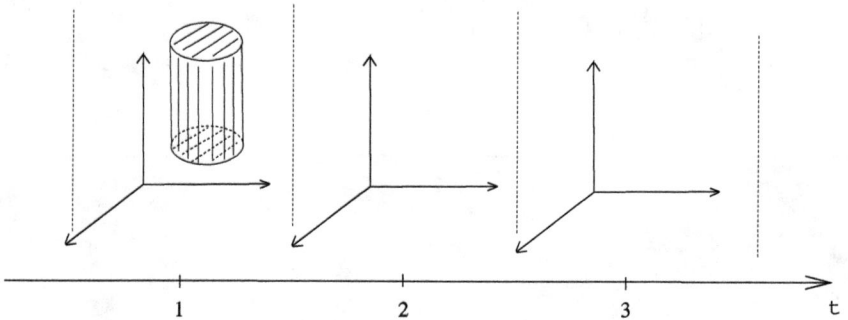

Figure 4.18.

Consider the objects shown at time $t = 1$ for the filled-in cylinder shown in Figure 4.16 and for the empty cylinder shown in Figure 4.18.

Note that these two objects are indistinguishable. It is what happens between times $t = 1$ and $t = 3$ that makes the difference. Use your imagination and compare them. Not only picture but also 'picture plus fantasisation' is important.

We are almost ready to answer the question we posed about the Hopf link (see Figure 4.19). The answer is the climax of the following section!

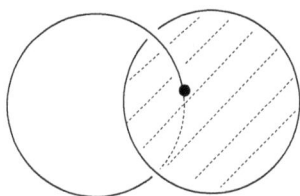

In $\mathbb{R}^4$ we can attach a disc to one circle in the Hopf link so that the disc does not touch the other circle, although we cannot do it in $\mathbb{R}^3$.

How do you do that?

Figure 4.19.   Question 4.1 on page 32. See also page 34

But, before we do that, let's consider one last example. Suppose that at time $t = 1$ second a circular light is shown and remains shining for two seconds. At precisely time $t = 3$ seconds, a disk-shaped light is shown instantaneously, where the boundary of this disk-shaped light is the same size as the original circular light. This process looks like what we illustrate in Figure 4.20. But what about in $\mathbb{R}^4$? What object is traced by the light?

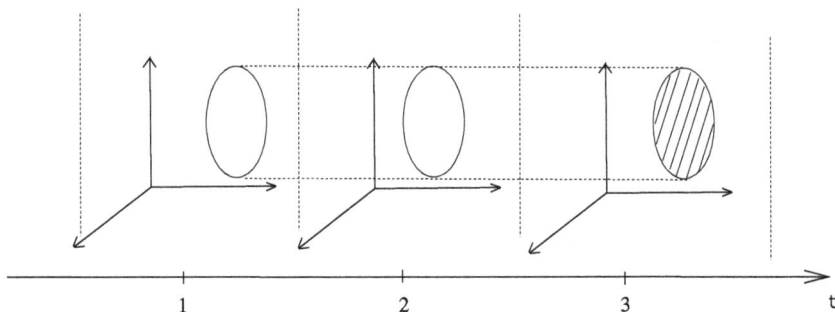

Figure 4.20.

The resultant object is a cylinder with only one face attached —
like a drinking glass with vertical sides. Imagine moving the object
shown in Figure 4.20 to the location shown in Figure 4.21 in $\mathbb{R}^4$.

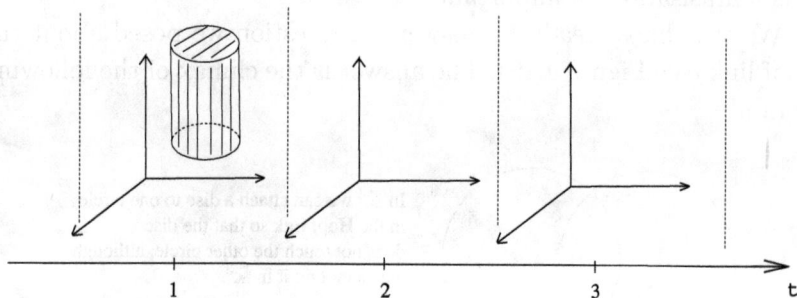

Figure 4.21.

By the way, we can mold this object continuously without cutting
or touching itself to obtain...

...a disk! Imagine the object to be made of clay. You can manipulate
it in the way shown in Figure 4.22.

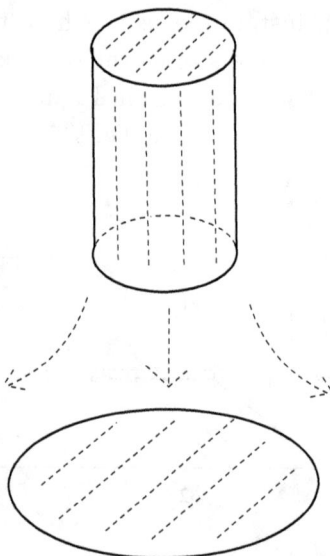

Figure 4.22.

We are now discussing our objects from the perspective that we can bend, stretch, and mold our object continuously without cutting or touching itself, as we please. Note that what transformation is allowed is defined rigorously in mathematics. One could imagine bending a plate up into the shape of a cup continuously without cutting or touching itself. In certain respects mathematically, these two objects can both be considered disks. Recall the discussion on page 34.

So, we can bend, stretch, and mold the object continuously without cutting or touching itself shown at $t = 1$ in Figure 4.21 to obtain the object shown at $t = 1$ in Figure 4.23.

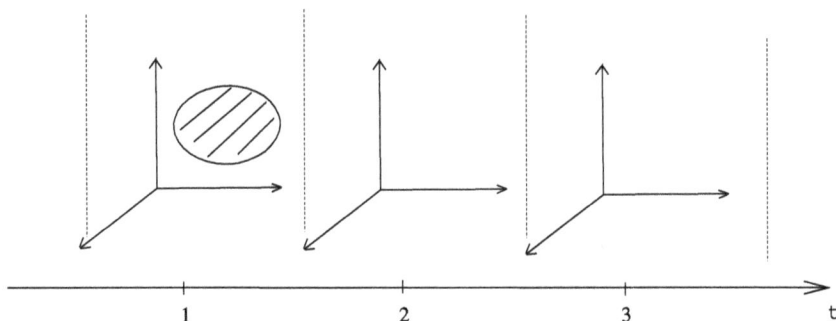

Figure 4.23.

In the following section, §4.5, we finally answer the question we posed about the Hopf link (see Figure 4.24).

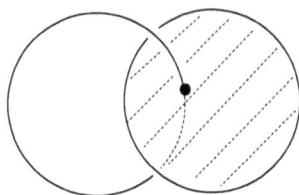

In $\mathbb{R}^4$ we can attach a disc to one circle in the Hopf link so that the disc does not touch the other circle, although we cannot do it in $\mathbb{R}^3$.

How do you do that?

Figure 4.24.   Question 4.1 on page 32. See also page 34

Are you ready for the answer? Have you already answered it yourself?

4.5. The Hopf Link and Four-Dimensional Space

We told the story on pages xxi and xxii to help the readers in this very moment — to provide clarity on the following explanation. See Figure 4.25. In $\mathbb{R}^2$, there is no way to connect a point A inside a circle (a closed curve) to a point B outside the circle via a curve without touching the circle. However,

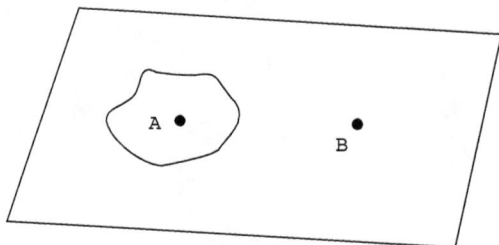

Figure 4.25.   The same figure as Figure 1 on page xxi

if we think $\mathbb{R}^2$ as a plane living in $\mathbb{R}^3$ (that is, think of $\mathbb{R}^2$ as drawn in Figure 4.25 as a subset of Figure 4.26), we totally can as drawn in Figure 4.26.

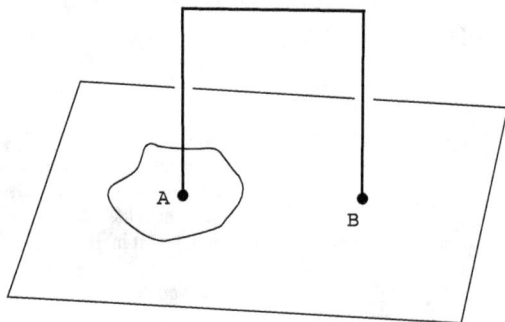

Figure 4.26.   The same figure as Figure 2 on page xxii

Now, then, let's go into four-dimensional space. How can we attach a disk to one component of the Hopf link without touching the other component? We are finally answering the question on th Hopf link shown in Figure 4.27.

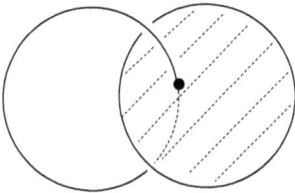

In $\mathbb{R}^4$ we can attach a disc to one circle in the Hopf link so that the disc does not touch the other circle, although we cannot do it in $\mathbb{R}^3$.

How do you do that?

Figure 4.27.   Question 4.1 on page 32. See also page 34

We do so in the way shown in a combination of Figures 4.28 and 4.29.

Specifically, take $\mathbb{R}^4$, where points are characterized by their width, depth, height, and time coordinates. Consider a Hopf link at time $t = 1$. One circle of the Hopf link exists only at time $t = 1$. However, we can decide that the other component exists for all t such that $1 \leq t \leq 3$. Insert a disk at $t = 3$ whose boundary is the component living at all such that $1 \leq t \leq 3$; suppose that this disk *only* exists at $t = 3$.

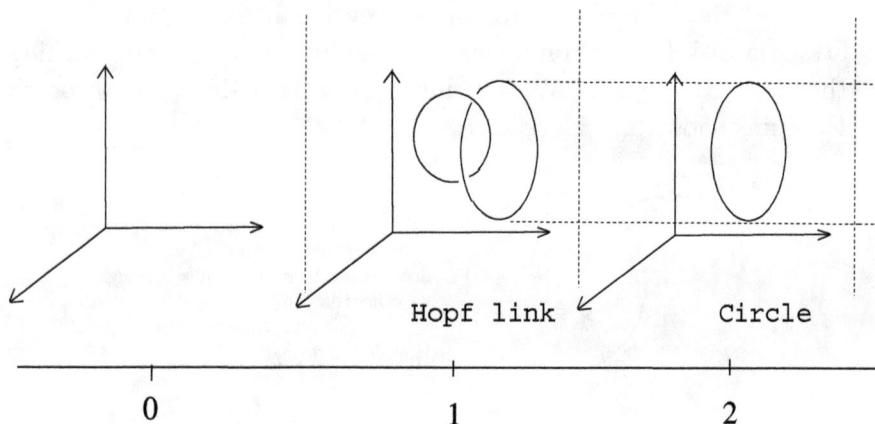

Figure 4.28.   Figures 4.28 and 4.29 make one figure

Now, look at all the objects in $\mathbb{R}^4$ in a combination of Figures 4.28 and 4.29, or Figure 4.30. We see that we successfully attached a disk to the right-hand component of the Hopf link without touching the left-hand component.

Draw as large a picture as you can, then you will imagine four-dimensional space better.

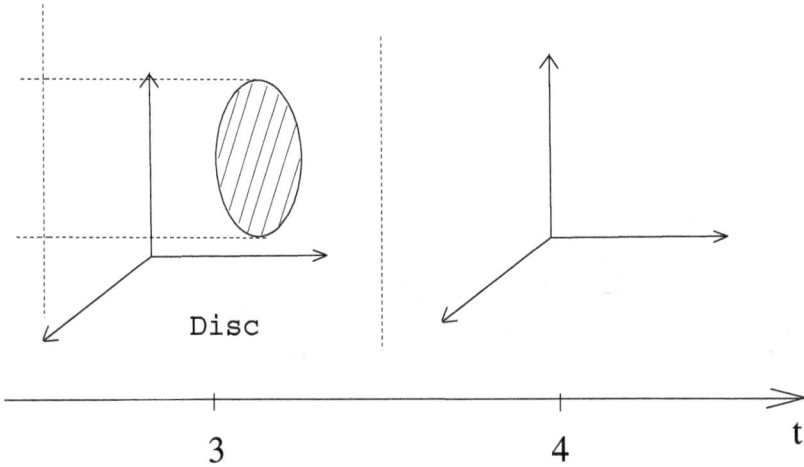

Figure 4.29. Figures 4.28 and 4.29 make one figure

We omit both sides of a combination of Figures 4.28 and 4.29, and draw it as in Figure 4.30.

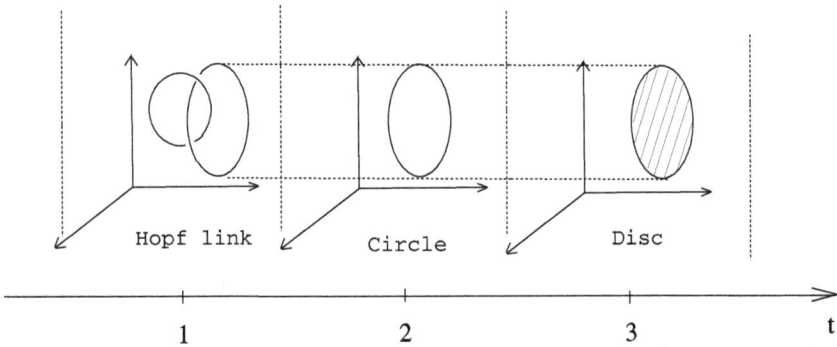

Figure 4.30.

Recall the discussion on and around page 34 about the drinking glass with perfectly vertical sides (a hollow cylinder with only one face). The method of attaching the disk to one component of the Hopf link makes use of this drinking glass-like object in four-dimensional space (see Figure 4.31). Note that its boundary in $\mathbb{R}^4$ is a circle at $t = 1$.

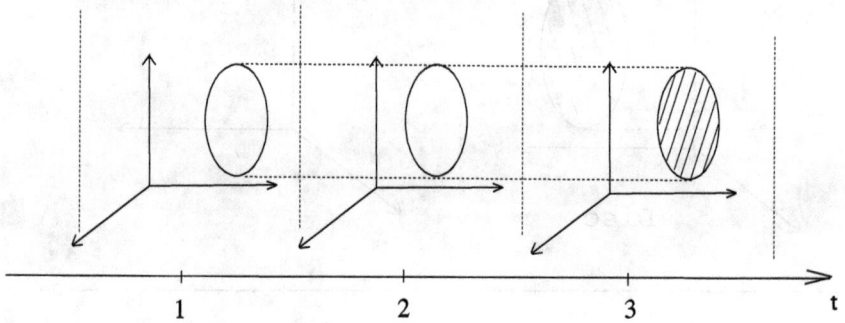

Figure 4.31.   The same figure as Figure 4.20

We can morph the drinking glass into a disk, as is done in Figure 4.32.

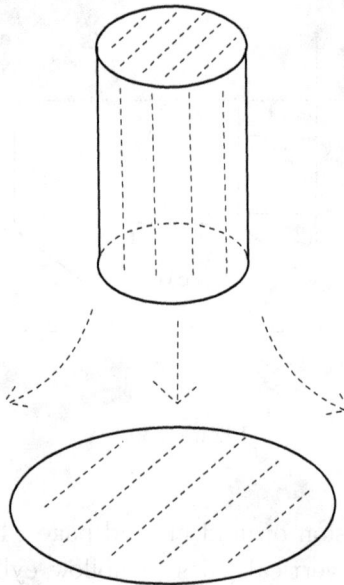

Figure 4.32.   The same figure as Figure 4.22

We remark that the Hopf link has two circles H and O. We attach a disk D only to H and that D does not touch O. Note that D is drawn in Figures 4.30 and 4.31. We show a sketchy proof that D does not touch O. Note that if they did touch, since O only exists at $t = 1$, D would have to touch O at this time coordinate. But 'a whole part of D at $t = 1$' is H. Hence, H and O touch. We arrived at a contradiction. Therefore, we confirm our success in attaching a disk to one component of the Hopf link without touching the other.

4.6.   A Construction using Paper, Wire, and Fire

In this section, we outline a simple paper construction which emulates the operation discussed in §4.5.

To start, you'll need a wire and a piece of paper. Bend the wire into a circle to make one component of the Hopf link and cut and/or tape the paper into a circle to form the second component, as is shown in Figure 4.33.

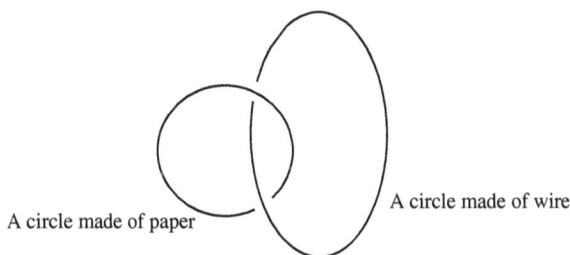

A circle made of paper   A circle made of wire

Figure 4.33.

Now, imagine what happens if we burn the paper circle. See Figure 4.34. This construction emulates what we see in $\mathbb{R}^3$ at and right after time $t = 1$; the components are linked precisely at $t = 1$ before one of the components (in this case, the paper one) vanishes.

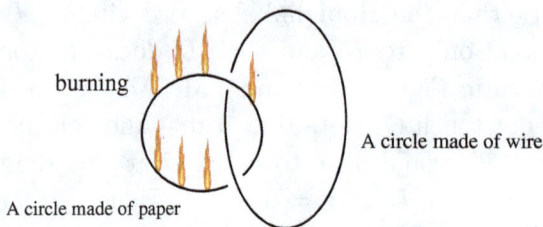

burning

A circle made of wire

A circle made of paper

Figure 4.34.

At the current time coordinate, the Hopf link is only a memory to us. Although, we must bear in mind that it still exists in $\mathbb{R}^4$ — just at a different time coordinate from the one we are currently at.

Note that after the paper circle burns up, the component crafted from wire still exists. This corresponds to what happens at times $1 \leq t \leq 3$, as is shown in Figure 4.35.

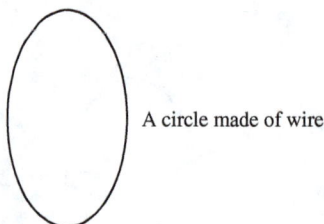

A circle made of wire

Figure 4.35.

Now, use another sheet of paper to cut out a disk that fits into the wire circle; attach this disk to the wire circle as in Figure 4.36. This process emulates what occurs at $t = 3$.

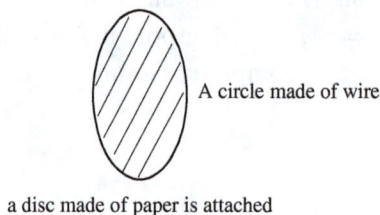

A circle made of wire

a disc made of paper is attached

Figure 4.36.

We need to use our imaginations for this paper construction, but that's okay! Imagination is often a very useful tool in understanding complicated mathematical constructions.

Miscellaneous: The Time Machine

Arriving at this section, the reader should be able to imagine some mathematical concepts in four-dimensional space. However, some confusion may linger — in particular, regarding the time coordinate axis.

It may be the case that you can envisage four-dimensional space just fine, yet it does not feel like a tangible concept for the following reason. If we can craft four-dimensional paper constructions in the world we live in using a special type of flashlight, or by performing the paper-wire-fire construction from Section 4.6, shouldn't we be able to time travel? If we live in $\mathbb{R}^4$, shouldn't we be able to travel into the past and future just like we can travel up and down, side to side?

In our examples, we moved freely along the time axis just as we would any of the physical positioning axes. We can imagine this process. But in the real world, we naturally move forward along the time axis; in order to move backwards along the time axis, we would have to invent a time machine. (Here, a time machine means a machine to move forward and backwards freely along the time axis.)

But can we really construct a time machine? At the time of writing of this book, as far as the author knows, it has not been invented. No one knows whether or not it is possible.

It would be a great scientific milestone if scientists were to develop a new mathematical theory or provide a physical explanation that proves the possibility or impossibility of time machine. Even mere evidence in either direction would be a great scientific achievement and a major historical landmark. So, why don't you give it a try? You could break ground in this fascinating but mysterious area of research. Excitement awaits you! Perhaps you will build the first time machine.

Chapter 5

The Klein Bottle

5.1. It Is Not a Torus

Consider a square $ABCD$ consisting of its boundary and the body and add arrows along the edges as shown in Figure 5.1.

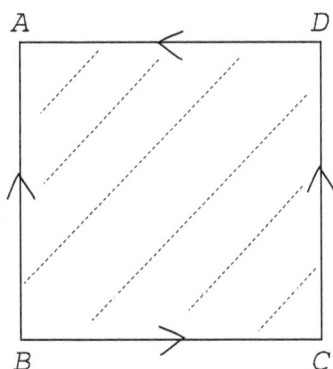

Figure 5.1. A square

Note that the directions of the arrows are different from those we used to construct the torus in Figure 5.2.

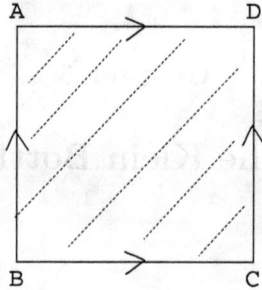

Figure 5.2.   The square in Figure 3.1 becomes a torus

Attach the segments *AB* and *DC* as in Figure 5.3 so that

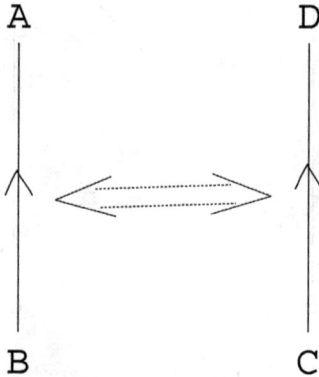

Figure 5.3.

the arrows line up. Note that according to this gluing, *A* and *B* become identified with one another, and *B* and *C* become identified with one another.

Attach the segments BC and AD as shown in Figure 5.4 so that

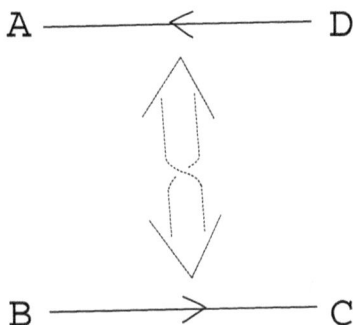

Figure 5.4.

the arrows line up. Now the points B and D become identified with one another, and A and C become identified with one another. Combined with our last gluing, we now have that the four points A, B, C, and D have all become identified as one point.

The reader may carry out both attachment on time, or separately in any order.

When performing this construction, you may take the liberty of bending and stretching the square without cutting itself in whatever way possible to make the edge gluings work. However, we are still to assume (just as we did when constructing the annulus, Möbius band, and torus) that no part of the square touches itself except for the edges identified in the gluing. We suppose that the resulting object does not touch itself. See page 13 to recall what we mean when we say an object "touches itself."

Question 5.1. Can we complete this operation? If so, what do we obtain? What shape does it have?

Go ahead — give it a try. First, make the identifications shown in Figure 5.5 to obtain an annulus.

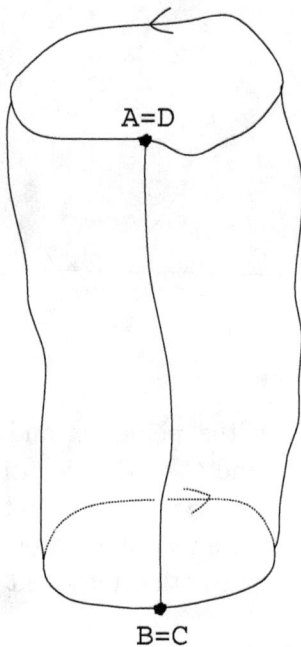

Figure 5.5.

Note the orientation of the arrows along DA and BC. Can you attach these segments so that the arrows coincide?

Does it seem impossible? Indeed, it is impossible.

Well, what if we try a different way?

Let's first identify DA and BC so that we do not run into the same issue as last time. The arrows are aligned in the same orientation as shown in Figure 5.6. Note that we have formed a Möbius strip.

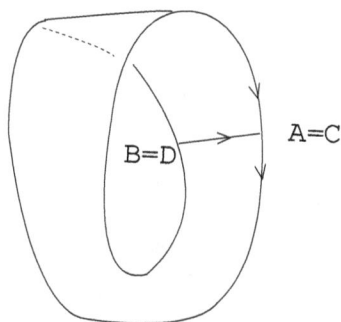

Figure 5.6.

Note in Figure 5.6 the orientations of the arrows along AB and DC. Can you attach these two segments so that their arrows coincide?

Does this also seem impossible? Alas, it *is* impossible. We cannot make this identification.

Is there any way that we can make both of the desired edge identifications? It certainly does not seem so. And in fact, it cannot be done. The gluing we have described by these arrows is not possible.

More specifically, it is known that the object of Question 5.1 cannot be constructed in three-dimensional space. This is precisely why we are having trouble constructing it from our gluing diagrams. Whichever way we try, the object must touch itself in order to complete the gluing.

However, you may have guessed what comes next.

While we cannot construct this object in three-dimensional space, we *can* construct it in four-dimensional space! We explain how to do so in the subsequent sections.

5.2. The Klein Bottle and $\mathbb{R}^3$

Recall the question we posed on page 57 (see also page 59); we rewrite it here in Figure 5.7.

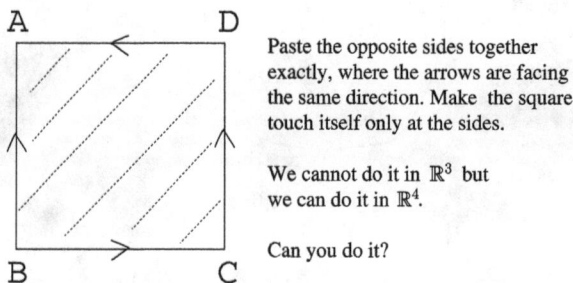

Paste the opposite sides together exactly, where the arrows are facing the same direction. Make the square touch itself only at the sides.

We cannot do it in $\mathbb{R}^3$ but we can do it in $\mathbb{R}^4$.

Can you do it?

Figure 5.7.   Question 5.1 on page 57. See also page 59

In the preceding section, we discussed how we cannot construct the desired object in $\mathbb{R}^3$, for when we try to make the second of two edge identifications, we end up forcing the object to touch itself. Or, even if we try to carry out both operations simultaneously, we cannot do that. We noted, however, that we can construct the object in $\mathbb{R}^4$.

When we appropriately identify the edges of our square as described in the last section, we obtain a valid object in $\mathbb{R}^4$. We call it the *Klein bottle*, named after a great mathematician, Felix Kelin (1849–1925). From what we can ascertain from the mathematical literature of that time, it seems that Klein was the first to explicitly discuss this object in great detail. Consequently, we have crowned the object with his name.

Let's imagine what shape the Klein bottle takes in $\mathbb{R}^4$. Fantasize extensively.

We prepare to do that by discussing another construction in $\mathbb{R}^3$. This alternative object is similar to the Klein bottle, though is subject to less strict criteria. Suppose that everything is the same as what we discussed in the previous section but that now we allow the object to intersect itself. This self-intersection is illustrated in the following (Figures 5.9 and 5.10).

Make the identifications shown in Figure 5.8.

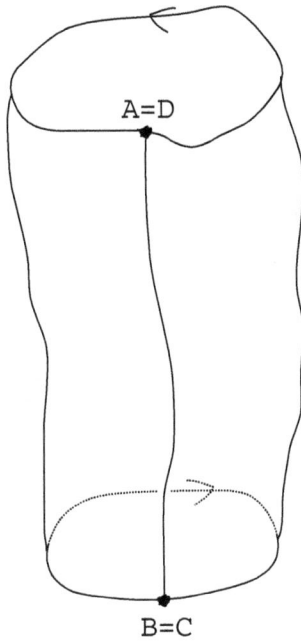

Figure 5.8.

Next, perform the operation shown in Figure 5.9.

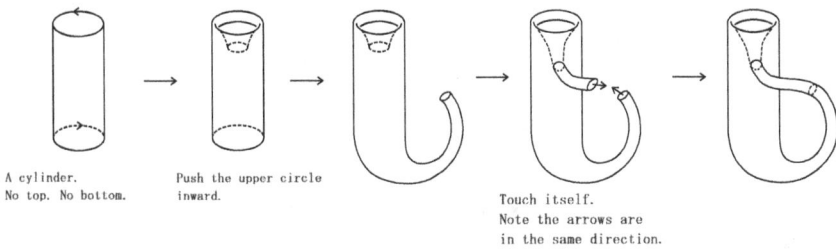

A cylinder.
No top. No bottom.

Push the upper circle
inward.

Touch itself.
Note the arrows are
in the same direction.

Figure 5.9.

We have now finished the construction. Note the self-intersection from the operation performed in Figure 5.9.

The rightmost object in Figure 5.9 (shown on its own in Figure 5.10) is what we call an *immersion* of the Klein bottle into $\mathbb{R}^3$.

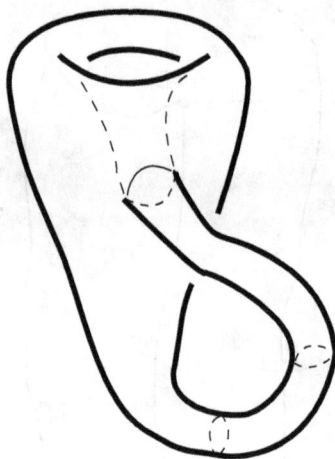

Figure 5.10.

Although we are unable to construct an object in its true form in a given space, we can at least still *immerse* the object in said space to get a sense of what it looks like. Note that the word, 'immerse', is defined mathematically rigorously.

Our immersion of the Klein bottle in $\mathbb{R}^3$ contains a pesky self-intersection. In $\mathbb{R}^4$, we can eliminate this self-intersection. Try to imagine this from what the object looks like immersed in $\mathbb{R}^3$ (Figure 5.10). Are you imagining it?

5.3.   The Klein Bottle and $\mathbb{R}^4$

We rewrite Question 5.1 on page 57 with comments on page 59 in Figure 5.11 to explain the construction outlined in the last part of §5.1.

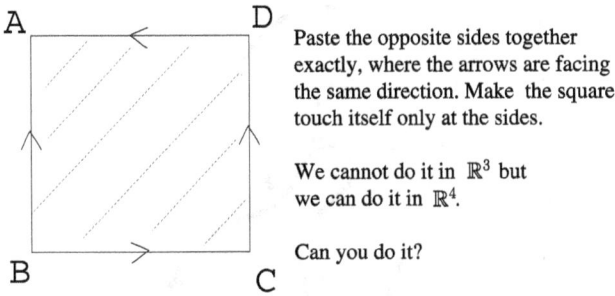

A D

Paste the opposite sides together exactly, where the arrows are facing the same direction. Make the square touch itself only at the sides.

We cannot do it in $\mathbb{R}^3$ but we can do it in $\mathbb{R}^4$.

Can you do it?

B C

Figure 5.11. Question 5.1 on page 57. See also page 59

Recall the constructions we did for the Hopf link (see Figure 5.12). In $\mathbb{R}^3$, we cannot fill one of the components with a solid disk without touching the other component. But, we saw in §4.5 that we *can* do so in $\mathbb{R}^4$. The case of the Klein bottle is an application and generalization of the Hopf link case.

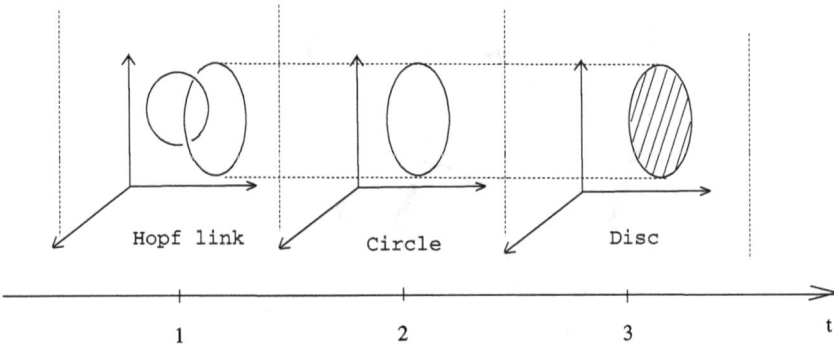

Hopf link   Circle   Disc

1   2   3   t

Figure 5.12. The same figure as Figure 4.30

Consider an immersion of the Klein bottle in $\mathbb{R}^3$ as shown in Figure 5.13.

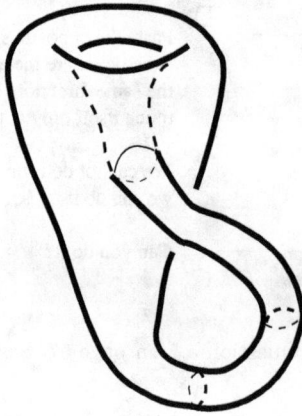

Figure 5.13.   The same figure as Figure 5.10

Now, drill a hole in it as is done in Figure 5.14.

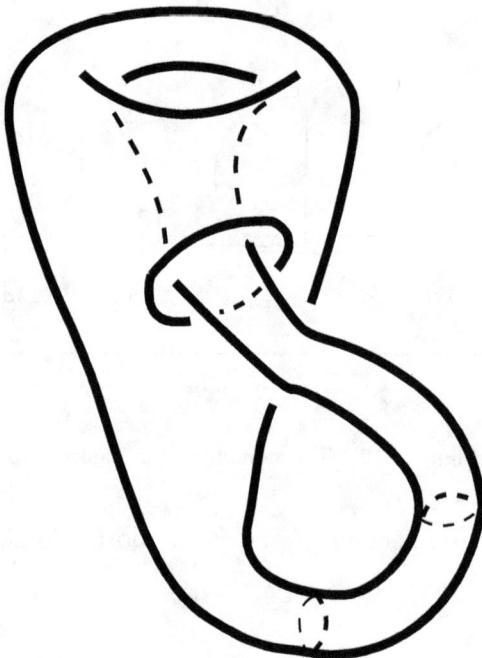

Figure 5.14.   A punctured Klein bottle

This object in Figure 5.14 is called a *punctured Klein bottle*; it is a Klein bottle missing a disk.

We make the Klein bottle in $\mathbb{R}^4$ similar to how we constructed the immersed Klein bottle in $\mathbb{R}^3$ (Figure 5.15).

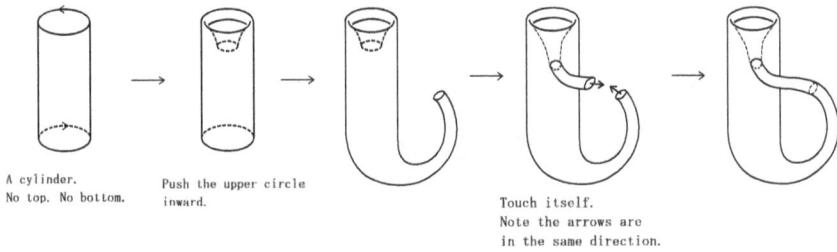

A cylinder.
No top. No bottom.

Push the upper circle inward.

Touch itself.
Note the arrows are
in the same direction.

Figure 5.15. The same figure as Figure 5.9

Continuing on, drill a hole in the cylinder on the far left of Figure 5.15. The procedure now mirrors that of constructing the immersion in $\mathbb{R}^3$. See Figure 5.16.

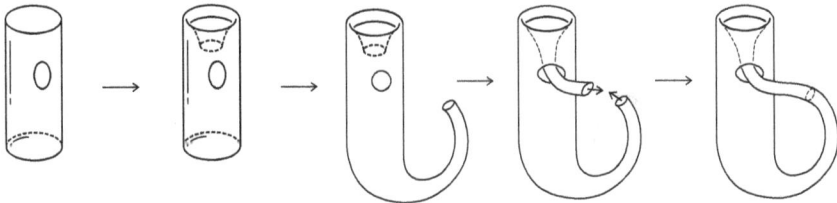

Figure 5.16.

Note that if we try to reattach a disk into the hole of a punctured Klein bottle, exhibited in Figure 5.14, in $\mathbb{R}^3$, we cannot help but force the object to once again touch itself. It is the same phenomenon as that pictured in Figure 5.13.

But we reiterate: In $\mathbb{R}^4$, we *can* reattach a disk into the drilled hole, all the while preventing any self-touching.

You may be thinking that we will use a similar method to the one we used to attach a disk to one component of the Hopf link (Figure 5.17).

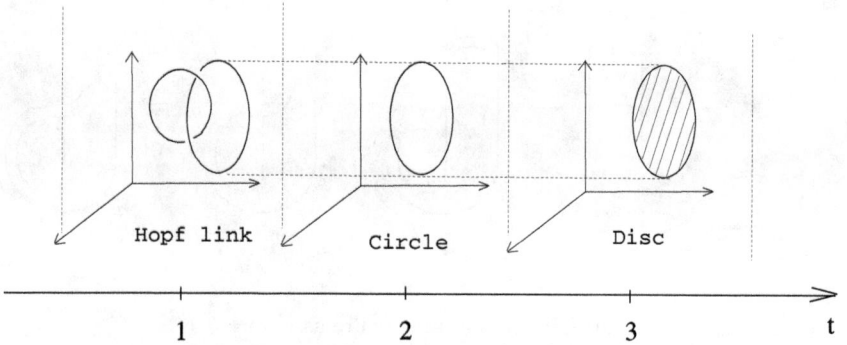

Figure 5.17.   The same figure as Figure 4.30

Note the following aspect of Figure 5.17 which is drawn in Figure 5.18.

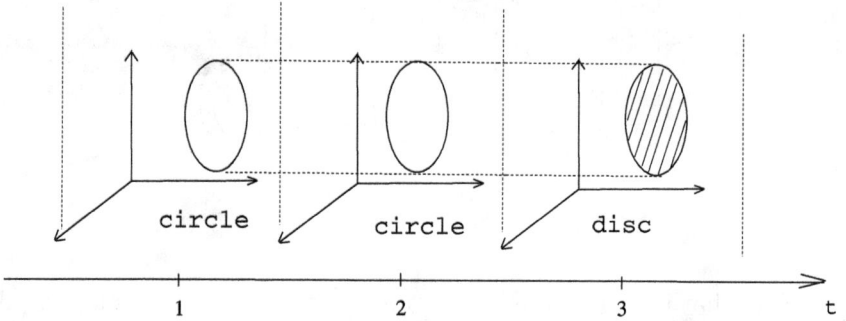

Figure 5.18.   Part of Figure 5.17, and Figure 4.20

Now, proceed as in a combination of Figures 5.19 and 5.20. Consider a point in $\mathbb{R}^4$ to be determined by width, depth, height, and time.

The punctured Klein bottle — whose boundary is a circle — exists in $\mathbb{R}^3$ at $t = 1$. We attach a disk to the boundary of the punctured Klein bottle as follows.

Suppose that the punctured Klein bottle — except for the boundary circle — exists only instantaneously. That is, suppose that the whole object except for the boundary appears at $t = 1$ and vanishes for $t \neq 1$. The circle boundary, however, exists for time $1 \leq t \leq 3$. Attach a disk to the circle in $\mathbb{R}^3$ at time $t = 3$, just as we did for the Hopf link.

Consider now the union of all circles during the interval $1 \leq t \leq 3$ with the disk at $t = 3$. This union forms the same object we discussed during our construction of the Hopf link with a filled in component in $\mathbb{R}^4$ (as in Figure 4.20 and in Figure 5.18).

This union can be molded continuously into a disk without cutting or touching itself just as before (Figure 4.23), so we may regard it as a disk. We have thus succeeded in attaching a disk into the hole in the punctured Klein bottle. Note that we have done so without forcing the object to touch itself. Consequently, we obtain a proper Klein bottle, as we have performed the proper edge identifications and the resultant object does not touch itself.

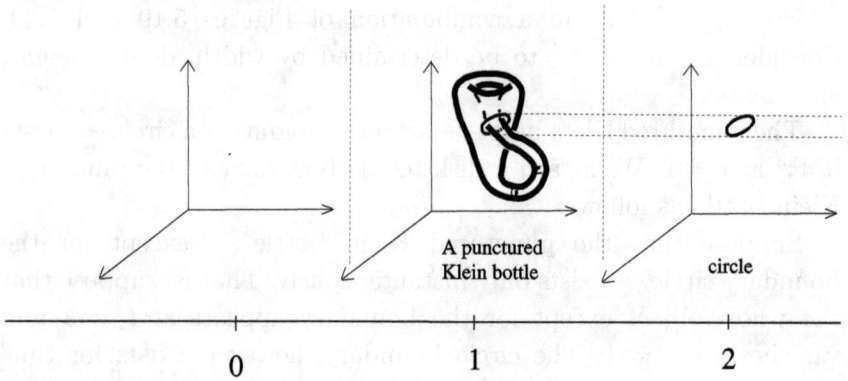

A punctured
Klein bottle

circle

0   1   2

Figure 5.19.   Figures 5.19 and 5.20 make one figure

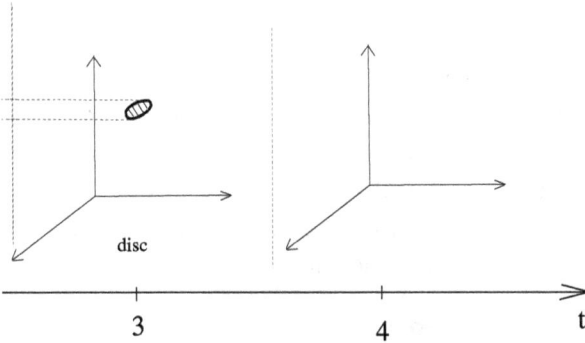

Figure 5.20.   Figures 5.19 and 5.20 make one figure

We omit both ends of a combination of Figures 5.19 and 5.20, and draw Figure 5.21.

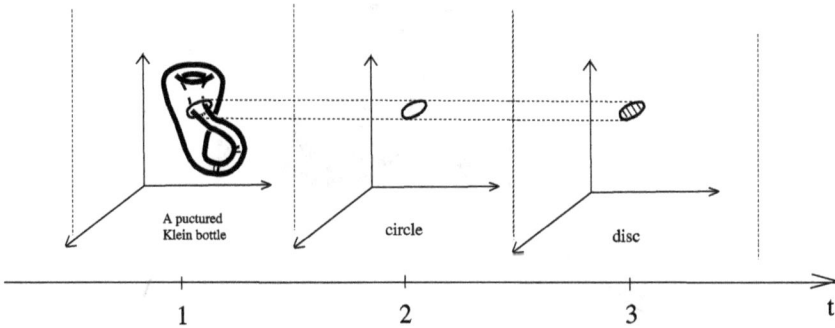

Figure 5.21.

We show a sketchy proof that the Klein bottle K in Figure 5.21 does not touch itself. It is enough to prove that there is not a self-touching in $\mathbb{R}^3$ at any t.

If $t < 1$ or $t > 3$, there is no part of K.
If $t < 1 < 3$, 'a whole part of K at t' is a circle and does not have a self-touching.
If $t = 3$, 'a whole part of K at $t = 3$' is a disc and does not have a self-touching.

If $t = 1$, 'a whole part of K at $t = 1$' is a punctured Klein bottle and does not have a self-touching.

Therefore, there does not exist a self-touching anywhere.

Note the dotted line in $\mathbb{R}^3$ at $t = 1$ in Figure 5.21; see Figure 5.22.

These dotted lines are meant to help the reader visualize where the circle and disk are at times $t = 2$ and $t = 3$. However, in $\mathbb{R}^3$ at $t = 1$, there is only the punctured Klein bottle; the dotted lines are purely conceptual and do not touch the the object.

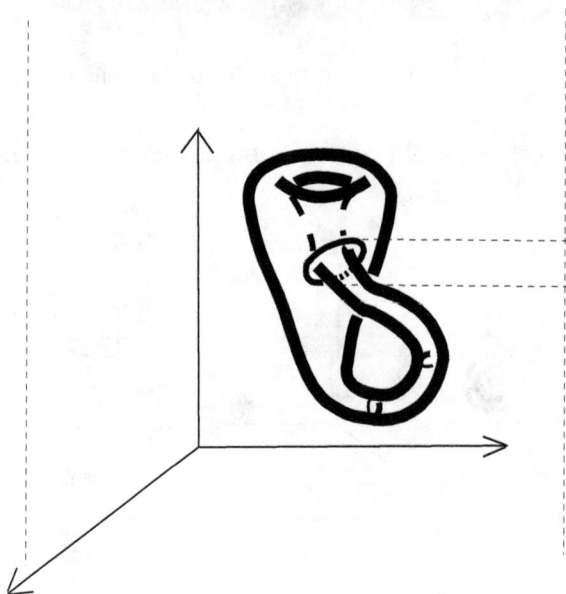

Figure 5.22. $\mathbb{R}^3$ at $t = 1$ in Figure 5.21

If we consider an immersion of the Klein bottle in $\mathbb{R}^3$, it always touches itself; the bottle neck must pass through the body in order to complete the gluing. However, we can easily construct a Klein bottle in $\mathbb{R}^4$ which does not touch itself by changing the time coordinates of certain parts of the object. The freedom given to us by moving into $\mathbb{R}^4$ captures the attention of any science fiction enthusiast and encourages people to write stories including Klein bottles and object with similar properties.

The author recalls a movie from his childhood where the main characters travel through a Klein bottle into a four-dimensional world. He has also read a comic where an accident in four-dimensional space cause the three-dimensional space in which we live to collapse! The Klein bottle is used to solve the catastrophe.

Just as the sci-fi characters do, the reader will now travel into four-dimensional space!

5.4. How to Construct Klein Bottle

Let's construct a Klein bottle. You may be wondering if we can use a stretchy, bendable material to craft our Klein bottle — or if such material is even available and affordable. Don't worry! Assume that we do have access to and can use a moldable material to construct our Klein bottle.

Using a PET plastic bottle, a pair of scissors, and a preferably thin and stretchy sock, we can construct a punctured Klein bottle shown in Figure 5.23, as well as

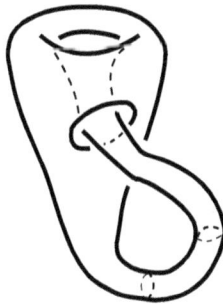

Figure 5.23.   The same figure as Figure 5.14

a Klein bottle immersed into $\mathbb{R}^3$, as shown in Figure 5.24.

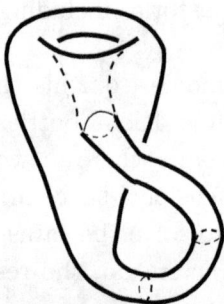

Figure 5.24. The same figure as Figure 5.10

The author made an instructional video for this DIY construction. He put it in YouTube. You can watch it by typing, "Ogasa Klein bottle" or 'Eiji Ogasa' into the search engine.

Prepare a transparent PET bottle, a pair of scissors, and a thin and easily stretchable sock, as in Figure 5.25.

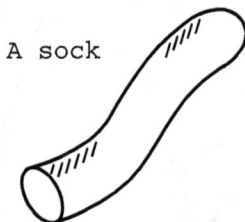

A sock

Figure 5.25.

See Figure 5.26. Using the scissors, cut off the bottom of the bottle as well as the spout and cut a hole into the side of the bottle. Cut off the tip of the sock; note that the resultant object is a transformed annulus.

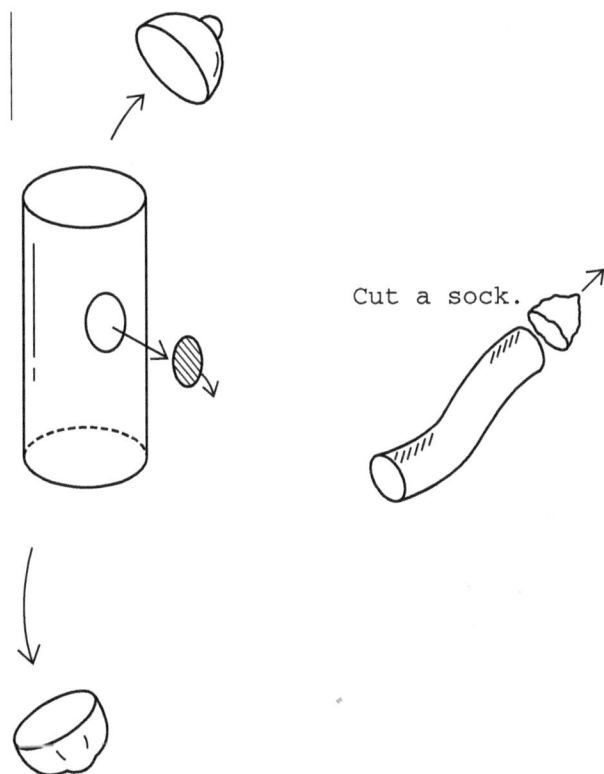

Cut a sock.

Figure 5.26.

The sock with a hole (transformed annulus) now appears as a big straw. Hook one end onto the circle formed by the bottom of the bottle. See the second left of Figure 5.27.

Next, pull the rest of the sock through the hole in the side of the bottle. Bring the other end of the sock up to the top of the bottle where the spout has been removed. See the second right of Figure 5.27.

Now pull the other end of the sock over the top of the plastic bottle. See the most right of Figure 5.27.

Thus, we made a punctured Klein bottle.

Pull the sock taut and cut as necessary to ensure that the sock does not touch itself.

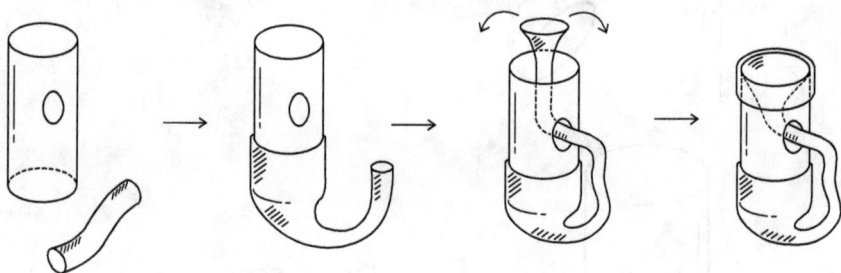

Figure 5.27.

We next construct an immersion of a Klein bottle in $\mathbb{R}^3$.

In the construction shown in Figure 5.26, we cut a hole in the side of the bottle by removing a slightly curved disk. Locate this disk and place it into the sock as shown on the right in Figure 5.28.

Once you have done this, you have successfully ascertained an immersion of a Klein bottle in $\mathbb{R}^3$.

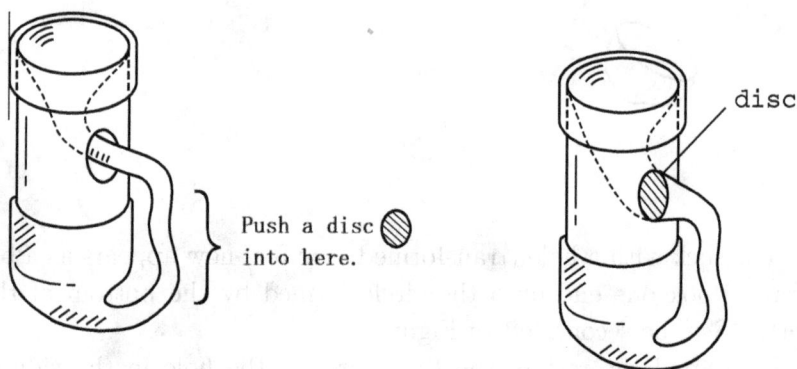

Figure 5.28.

We can imagine the disk existing at the same position but at a different point in time. And just like that, we have our Klein bottle in $\mathbb{R}^4$ (Recall Figures 5.19–5.21). Fantasisation is usually important in mathematics and physics.

5.5. Constructing the Torus

Next, we construct a torus. We begin the demonstration from the stage of bending a cylinder to glue up the sides. The reader knows already how to achieve the cylinder already from previous demonstrations.

You will need a PET plastic bottle, a pair of scissors, and another, thin, stretchy sock as in Figure 5.29.

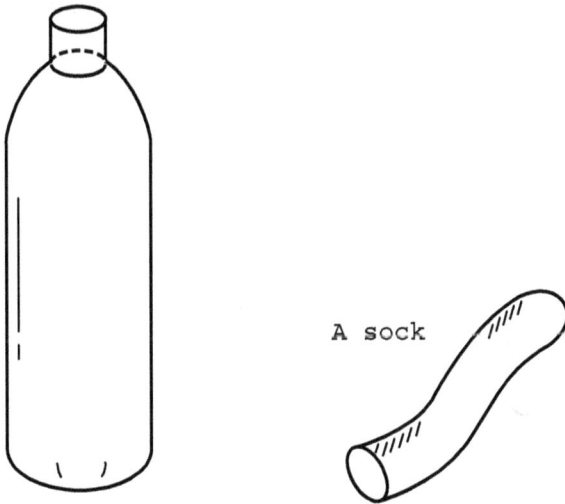

Figure 5.29.

As shown in Figure 5.30, cut off the bottom and spout of the bottle. Note that the resultant object is a transformed annulus. Now do the same to the sock as in the previous construction (Figure 5.26) — cut off the tip to get another transformed annulus.

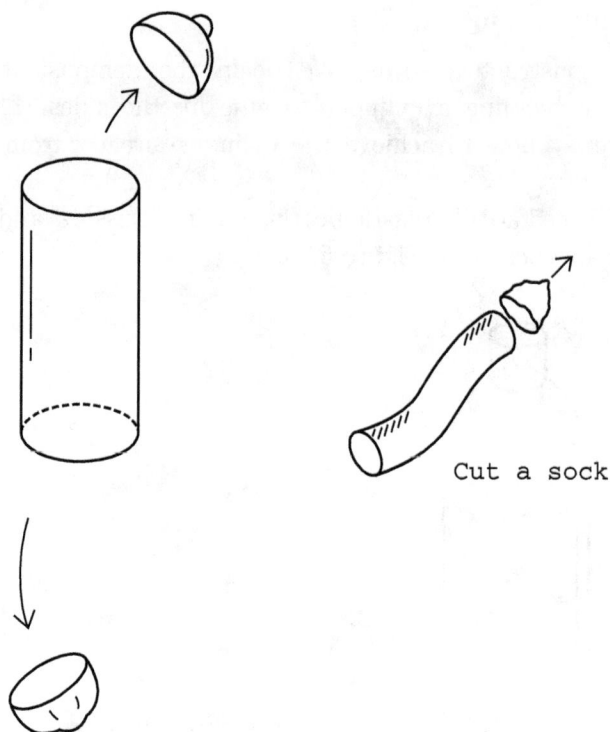

Cut a sock

Figure 5.30.

See Figure 5.31 for reference. Hook one end of the sock onto the circle created from removing the bottom of the bottle as shown in Figure 5.31.

Note that this object (a cylinder) is what we would achieve from gluing the vertical sides of a square (both oriented the same way).

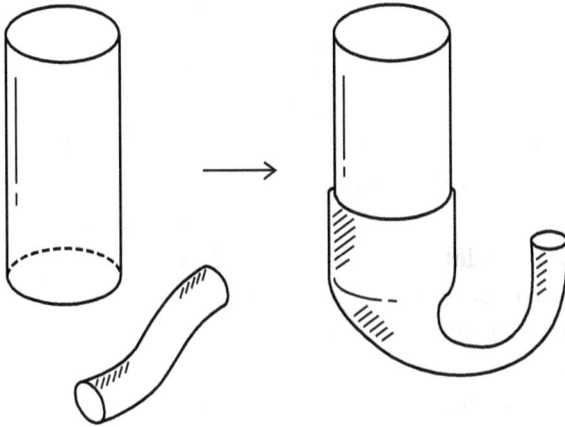

Figure 5.31.

See Figure 5.32. Take the other end of the sock and hook it over the top of the bottle.

Alter the size and tension in the sock as necessary to prevent it from touching itself. This completes the DIY construction of a torus, or at least in your imagination. Imagination is always important for these types of constructions.

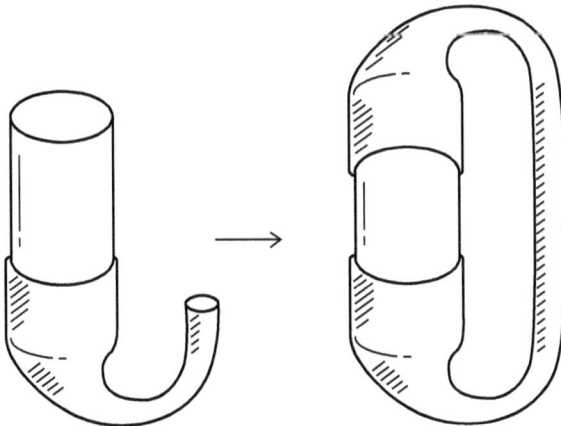

Figure 5.32.

Miscellaneous: The Theory of Relativity

There is an area of theoretical physics that discusses the relationship between time and space. This is called the *theory of relativity*. Throughout the course of this book, you learn about four-dimensional space $\mathbb{R}^4$ — the space determined by width, depth, height, and time. Understanding $\mathbb{R}^4$ will help you understand the theory of relativity.

The theory of relativity tells us the following about time machines. If you travel to space and fly at a very high velocity very far away from the earth for only one year, when you return, hundreds — even thousands — of years could have passed on earth. However, only one year has passed in your spaceship. You leave in the present and arrive well into the future. There are countless experiments to confirm this phenomenon.

Some readers may be convinced that this is in fact time travel. You would not necessarily be wrong in saying so, but there is no confirmation that it is possible to travel backwards along the time axis. So, it's still possible that a time machine that can go not only forward in time but also backward could still be invented by you!

Chapter 6

Möbius Band + Möbius Band = Klein Bottle

6.1. One or Two Sides

The Möbius band and an annulus are drawn in Figure 6.1. They look similar but have some fundamentally different properties. We said some of them in §2.1 and §2.2.

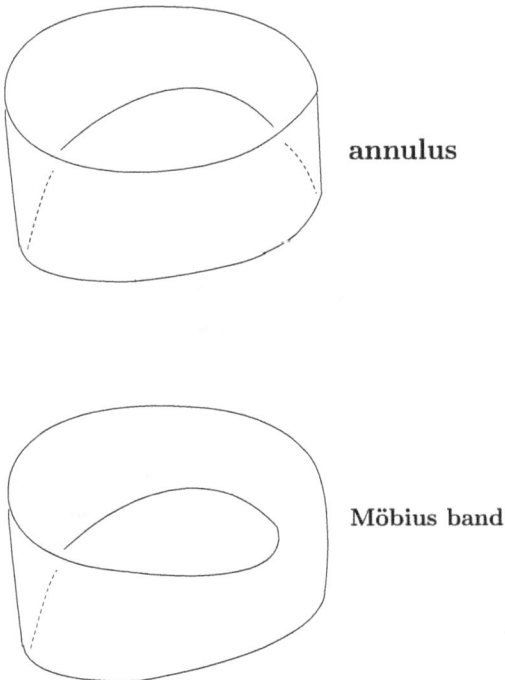

Figure 6.1. Möbius band and an annulus

Make the Möbius band and an annulus from a piece of paper. Use a pen, e.g. a ballpoint pen, that ink on one side does not soak through to the other side.

Draw a color at only a small part of Möbius band (respectively, an annulus) as drawn in Figure 6.2. Only one side is drawn by the color. Note that we use such a pen and a piece of paper.

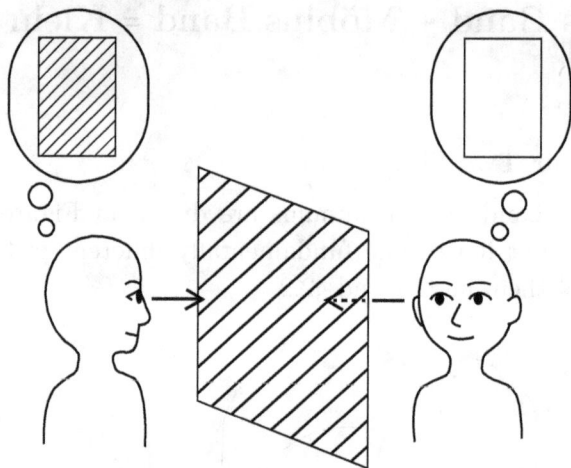

Figure 6.2.   A piece of paper. Use a ballpoint pen such that ink on one side does not soak through to the other side

For both the annulus and the Möbius band, continue to color the side you started coloring (but don't continue coloring over an edge). What happens in each case?

An annulus is colored only one side. The other side is not colored. See Figure 6.3. How about the case of Möbius band?

Do you feel that it is impossible to color only one side and not to color the other? Indeed, it is known that it cannot be done.

We say that Möbius band has only one side, as opposed to the annulus, which has two sides. This is one important difference between the annulus and the Möbius band.

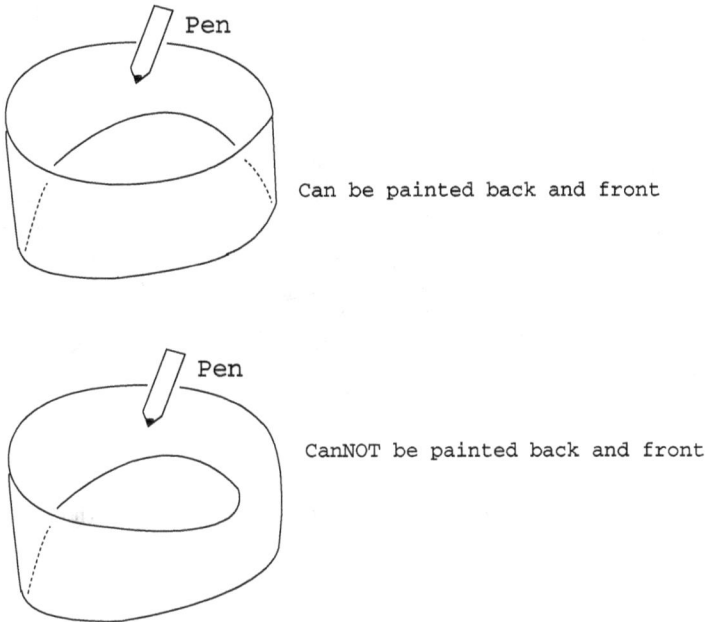

Figure 6.3.

6.2. Non-orientability

Next, use a pen and a piece of paper that allows ink to soak through to the reverse side. For example, use a paper towel used in kitchen.

Draw a round arrow on it by ink. Two persons, P and Q, look at the arrow in opposite directions, as shown in Figure 6.4.

Both people see the arrow go from A to B. However, person P sees it go around to the right while person Q sees the arrow go around to the left.

Draw a round arrow on the annulus, as in the upper one of Figure 6.5. Draw round arrows whose directions are the same as the one next to it as in the lower one of of Figure 6.5.

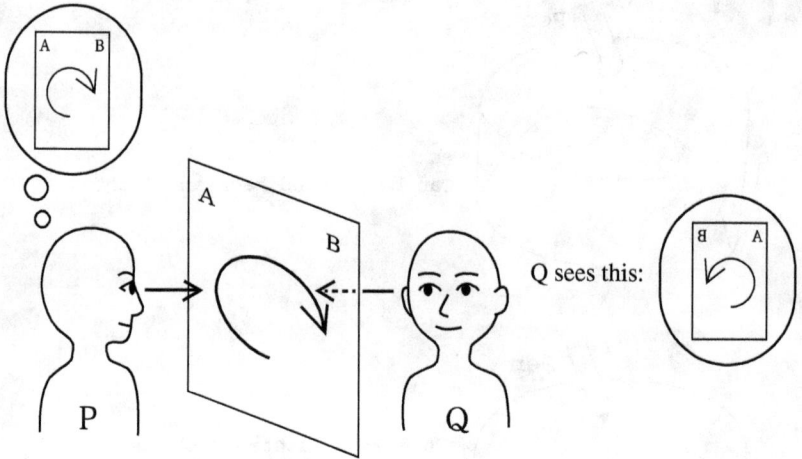

Figure 6.4.   A piece of paper that allows ink to soak through to the reverse side

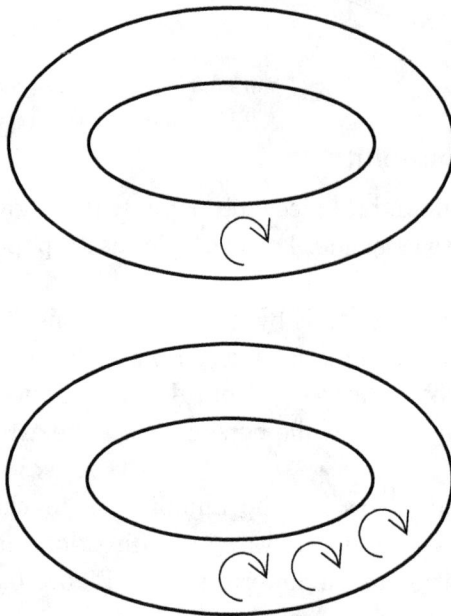

Figure 6.5.

As shown in Figure 6.6, draw round arrows so that the round arrows arrive at the starting round arrow.

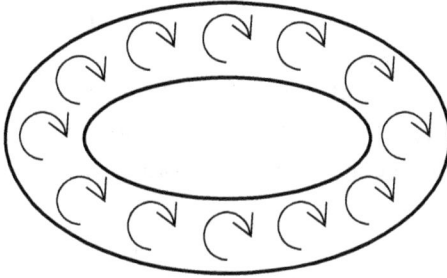

Figure 6.6.

Although you probably feel that it is obvious, a round arrow that arrived at the first round arrow has the same direction as that of the first round arrow.

How about it in the case of Möbius band? Do you think something different happens?

On Möbius band, draw a round arrow as in the upper one of Figure 6.7. Draw round arrows whose directions are the same as the one next to it, as in the lower one of Figure 6.7.

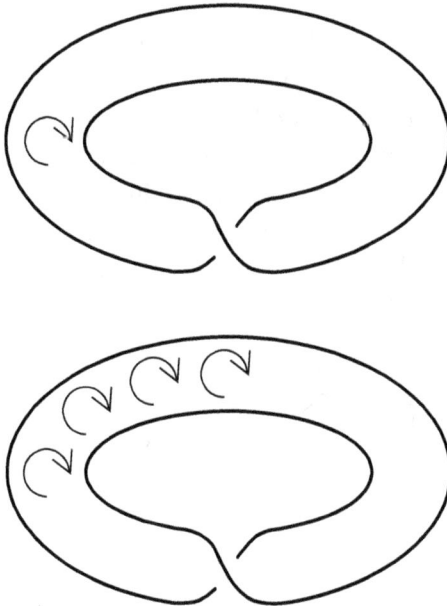

Figure 6.7.

Draw round arrows until a round arrow arrives at the first round arrow. What happens?

The round arrow that arrived at the first round arrow has the opposite direction to that of the first round arrow. See Figure 6.8.

This phenomenon is different from the case of an annulus (Figure 6.6).

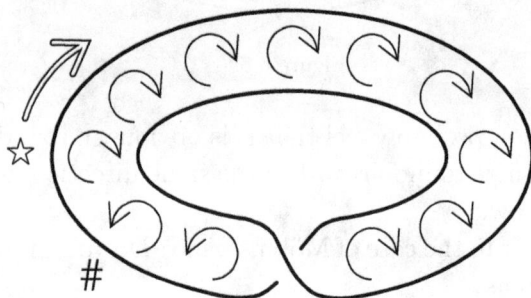

Start drawing ↻ from ☆ into the dirction ↗ .

Coming back to # , it is ↺ . It revereses.

Figure 6.8.   Use paper where the ink stains all the way to the back

As shown in Figure 6.9, we have a different situation in the case of the Möbius band from in the case of the annulus.

The orientation of round arrows is important. Don't take care where an open part of a round arrow is, or where the arrow draws on a round arrow. Both arrows drawn below are regarded as the same one.

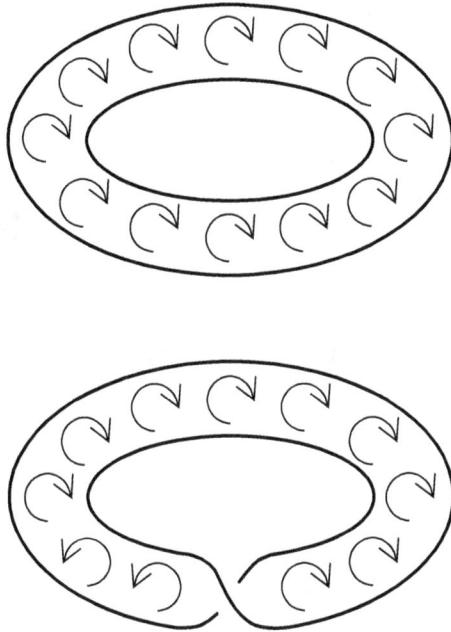

Figure 6.9.   Different phenomena

Some readers may feel it a little strange, but it is true.

We say that an annulus is *orientable* and that a Möbius band is *non orientable*.

We can align the directions of arrows in a small part, but we cannot align those of round arrows on the whole Möbius band.

Check it on your own by creating a Möbius band with a piece of paper that ink soaks from one side to the other, e.g. a paper towel used in the kitchen.

6.3.   Caution!

Some readers asked the author the following question. They tried drawing round arrows on Möbius band as in §6.2. They said that they 'succeeded' to align the directions of all round arrows as in

Figure 6.10. The readers wondered why. Do you understand the reason?

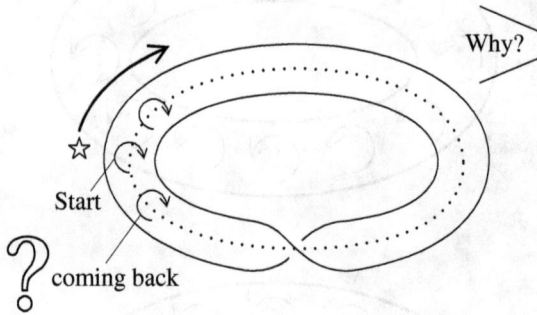

Figure 6.10.   What happens?

The reason is as follows.

In §6.2, we use a piece of paper such that ink soaks from one side to the other. The reader who asked the above question used a pen such that ink does not soak to the back of a piece of paper.

Try to construct the situation by first creating a Möbius band.

Draw a round arrow on the Möbius band. Then we have the situation as in Figure 6.11.

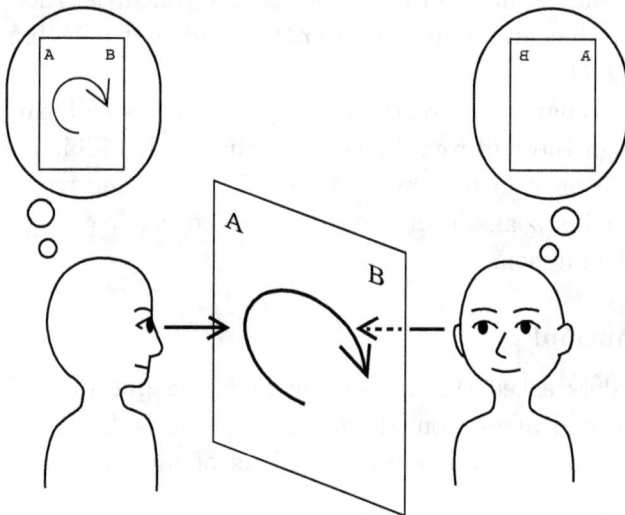

Figure 6.11.   Use a pen such that ink does not soak to the back of a piece of paper

As in Figure 6.12, draw round arrows so that one round arrow has the same direction as that of the one next to it. Near ◊, a round arrow is drawn on the 'opposite side'. Continue to draw round arrows. Near ☆, round arrows are drawn on the both sides. Continue drawing

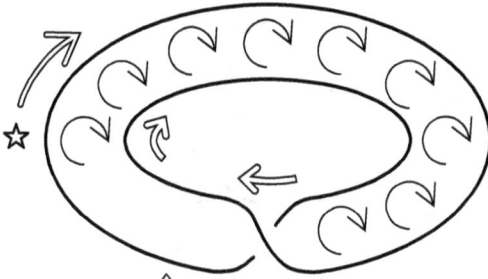

Start drawing ⟳
from ☆ ,
into the direction ⬈.
Coming back to ☆,
it is drawn on the opposite side.

At ☆, the situation is as follows

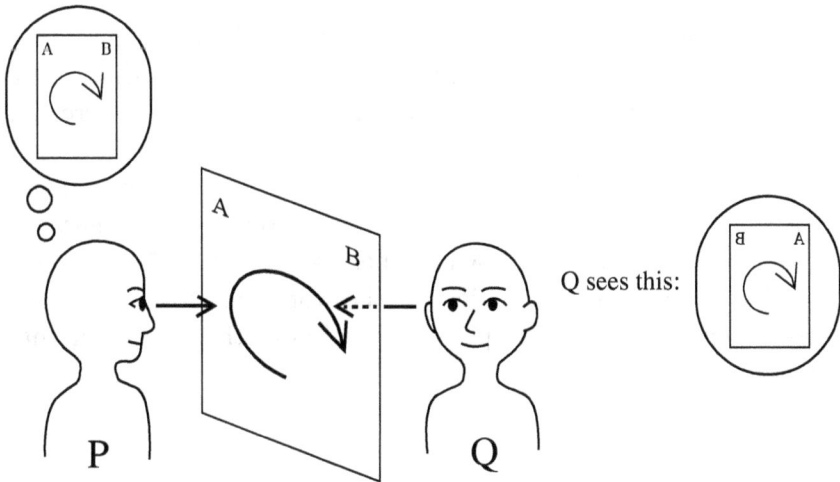

Q sees this:

Figure 6.12. Use a pen such that ink does not soak to the back of a piece of paper

round arrows in this way until you have covered the entire Möbius band.

A round arrow comes back to the place of the first round arrow as in Figure 6.13: Compare the round arrow near ♯ with that near ☆.

Figure 6.13.   Use a pen such that ink does not soak to the back of a piece of paper

In this case, while you draw round arrows, you go around Möbius band 'twice'. In Section 6.2 you go round only once.

That's a reason why the reader who asked the above question had a different result.

Recall Figure 6.8. On Möbius band, the round arrow that arrived at the first round arrow has the opposite direction to that of the first round arrow. This fact makes it a particularly interesting object for sci-fi writers. For example, if a character walked once around the center of the Möbius band, they would return to the same physical place. This could be portrayed by the character arriving at a parallel world where their world is turned inside out or right-left changing, an object in the world physically switching sides, or even reversing the personality of any other character than the hero.

The author watched a sci-fi drama where the Möbius strip was used as a means of transportation to travel into a different dimension.

Have you seen or read such stories?

6.4. Boundary

Next, we explore another difference between the Möbius band and an annulus. Use a pen and a piece of paper that allows ink to soak through to the reverse side.

From a point in the boundary of an annulus (respectively, Möbius band), begin to color along the boundary. Note that the ink should soak through to the reverse side as in Figure 6.14.

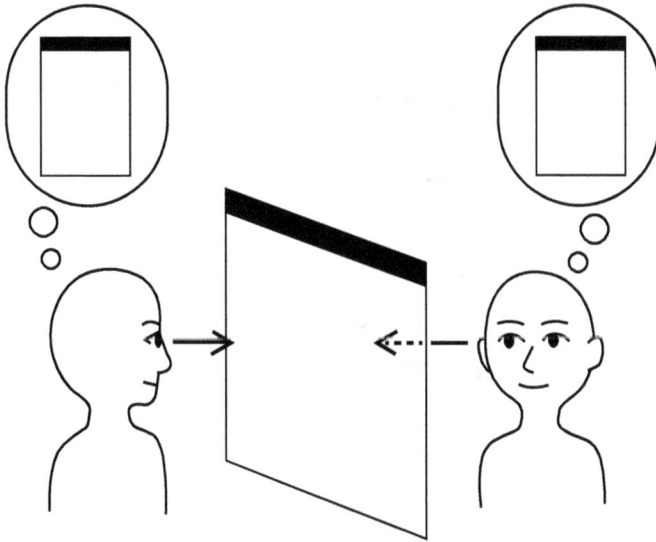

Figure 6.14. A piece of paper that allows ink to soak through to the reverse side

Do you note a difference between both?

The boundary of an annulus is a disjoint union of two circles while that of a Möbius band is only one circle, as shown in Figure 6.15.

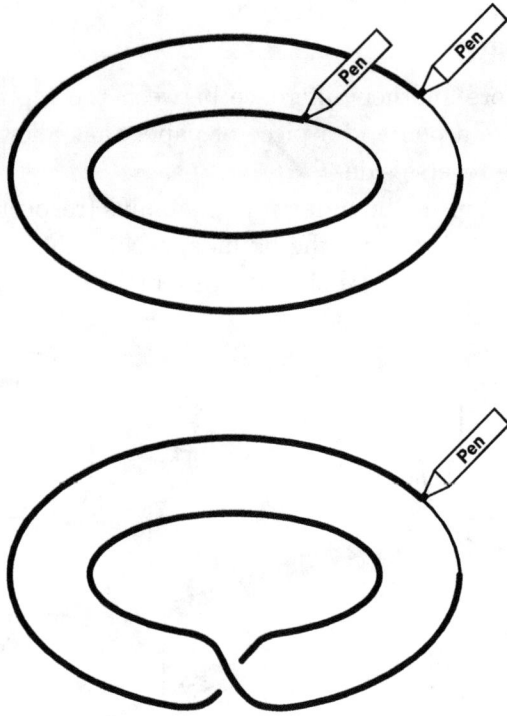

Figure 6.15.

This is an important difference between the Möbius band and an annulus.

6.5.   Klein Bottle = 2× (Möbius Band)

The Klein bottle can be divided into two Möbius bands, and conversely, two Möbius bands can be glued together to make one Klein bottle.

Recall that a Möbius band appears when we constructed a Klein bottle on page 59. This might lead you to guess at a connection between these two objects.

We introduce a DIY construction of Klein bottle = 2× (Möbius band). The author made a movie to demonstrate this DIY construction. He put it in YouTube. You can also find it by typing in the author's name, 'Eiji Ogasa', or 'Ogasa Klein bottle'.

Prepare a transparent PET bottle, a pair of scissors, and a semi-translucent sock (thin and easily stretchable), as in Figure 6.16. It is important that the sock are semi-transparent.

When we made Klein bottle page 72, we may use a non-transparent sock in Figure 5.25. However, in this case, you must use a semi-translucent sock.

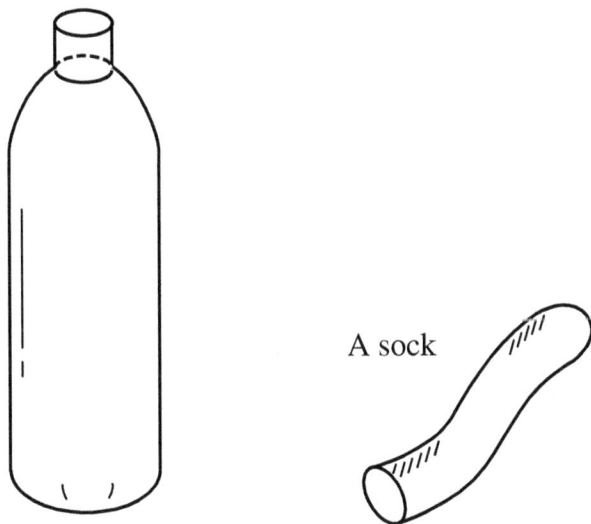

A sock

Figure 6.16. We must use a semi-translucent sock now although in Figure 5.25 we may use a non-transparent sock

We construct the Klein bottle as in Figure 6.17. Cut the top and bottom of the PET bottle. Cut a hole in the side of the PET bottle. Cut off the tips of the sock. (It is now an annulus.)

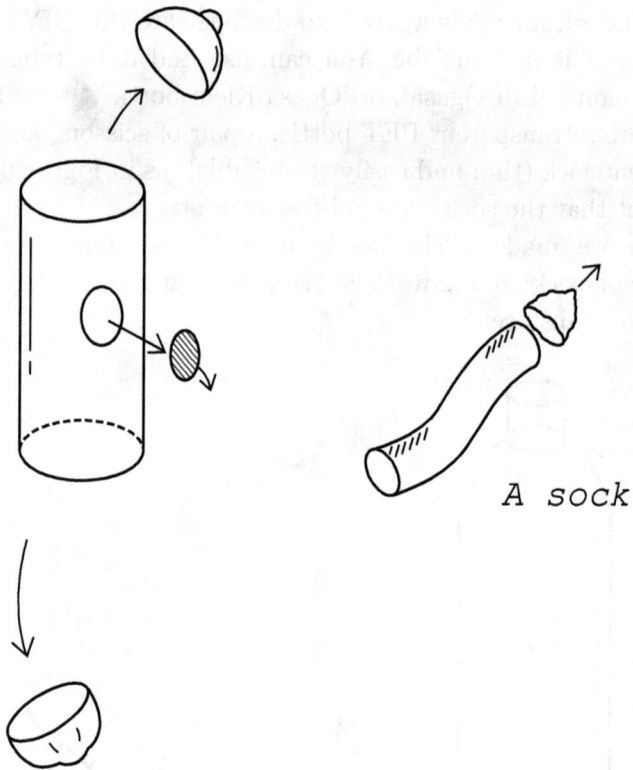

A sock

Figure 6.17.

Paint half of the PET bottle (respectively, the sock) red, as shown in Figure 6.18.

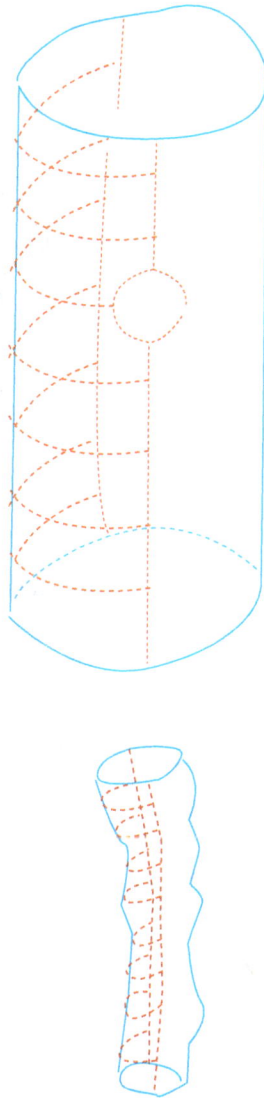

Figure 6.18.

Here, the fact that they're translucent helps.

Craft the Klein bottle minus a disc as we did in Figure 6.19.

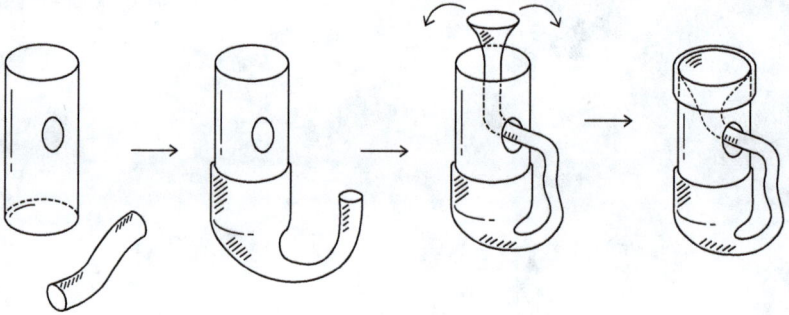

Figure 6.19.   The same figure as Figure 5.27

Then the red part and the semi-translucent part will look like the situation in Figure 6.20.

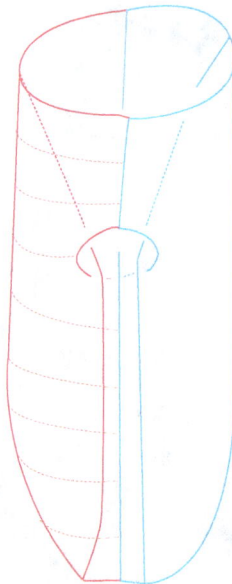

Figure 6.20.

Cut the object in Figure 6.20 with a pair of scissors along the border between the red part and the semi-translucent part.

As shown in Figure 6.21, the result is split into two parts. Did you note that both parts are the Möbius band? Did you note that they twist in the 'opposite' way?

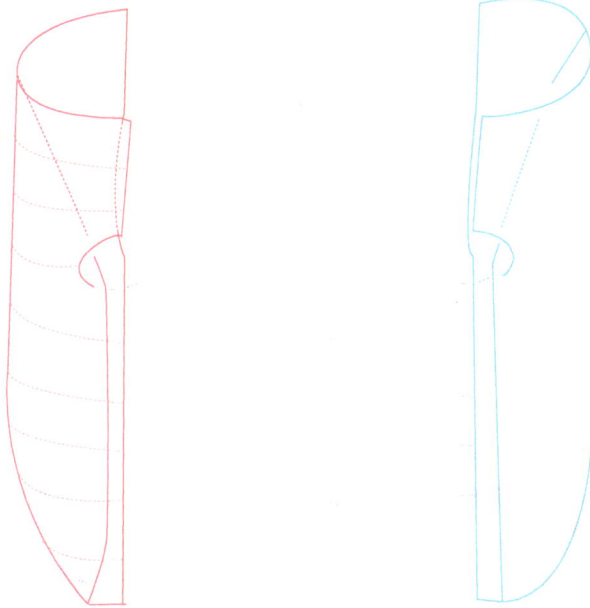

Figure 6.21. Caution! Both are the Möbius band

Take the disc that is cut from a plastic bottle when we make the object in Figure 6.20. Paint only half of the disk red. Recall how we put the disk in Figure 5.28. Place this disc as in Figure 6.22. Then an immersed Klein bottle into a three-dimensional space $\mathbb{R}^3$ is made from two Möbius bands.

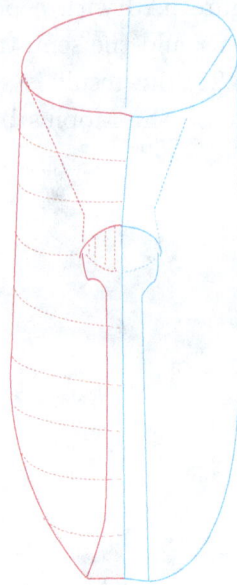

Figure 6.22.   A disc is placed as it is not in Figure 6.20

Attach the disk with only half-painted red to a punctured Klein bottle in $\mathbb{R}^4$, as shown in Figure 6.23. We see that the Klein bottle is made from two Möbius bands in four-dimensional space.

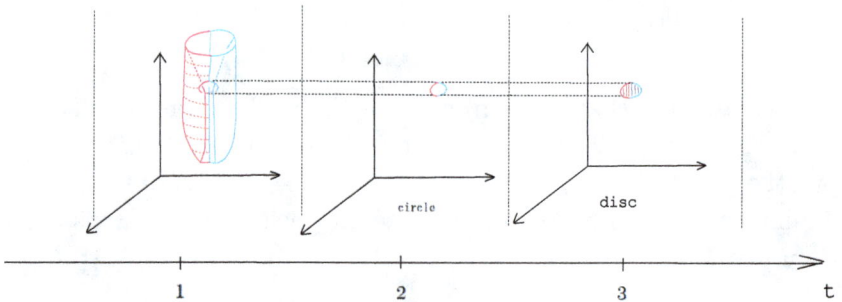

circle

disc

1   2   3   t

Figure 6.23.

Recall the following facts on page 85: an annulus is orientable. The Möbius band is non-orientable.

Look at a small part of Klein bottle. The part is orientable. (Recall the method using round arrows in §6.2.) Now, is the Klein bottle orientable?

The answer is negative. *Reason:* We use reductio *ad absurdum.* The Klein bottle contains a Möbius band, as we saw in this section. If Klein bottle is orientable, then the Möbius band must be orientable. We arrived at a contradiction. Therefore, our assumption that the Klein bottle was orientable must be false, and we conclude that the Klein bottle is non-orientable.

This is one of the important properties of Klein bottle.

Miscellaneous: Quantum Mechanics

In addition to the theory of relativity, quantum mechanics also involves research about time and space in the real world.

You might be interested to learn more about quantum mechanics on your own — this is just a brief introduction!

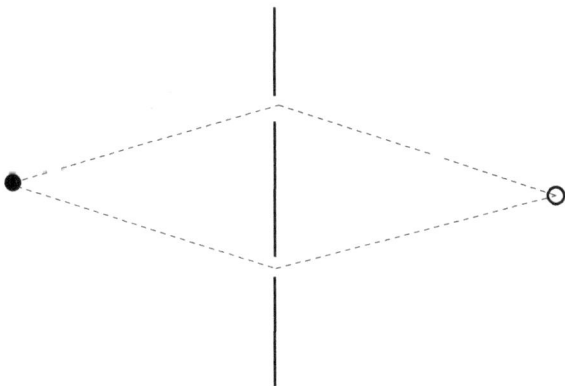

Figure 6.24.

See Figure 6.24. When the light, or whatever 'thing', went from ● to ○, we really cannot say whether it went through the top hole or the bottom hole. No matter how rigorous the experiments are, we will never know. In general, we can only say the probability that the object passed a certain way is about this much. This uncertainty is the foundation of quantum mechanics.

Some readers may want to know the reason why our nature has this intrinsic uncertainty. However, we accept simply this uncertainty and we do not ask why.

If you accept that is simply the way the world works, the theory of quantum mechanics will accurately predict the probability of the results of the experiment. For some of you, this accuracy of our predictions is enough.

Some of you might feel differently. The very first person who discovered quantum mechanics could be happy about discovering something new about how our world works. But you might wonder *why* this is true instead of simply accepting that the theory accurately predicts experimental results.

Well, maybe it's just human nature. To get rid of that feeling, what should you do? You have to discover something new on your own. That's really true. You will discover and invent something new!

Chapter 7

Boy Surface

7.1. It Is Not the Torus or Klein Bottle

Prepare a square as shown in Figure 7.1, making sure to include the arrows on the edges.

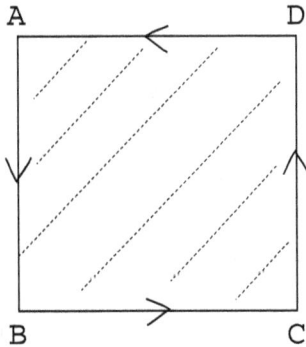

Figure 7.1.   A square

Note that the directions of arrows in Figure 7.1 are different from the case of a torus and that of Klein bottle. See Figures 7.2 and 7.3.

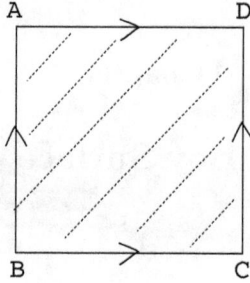

Figure 7.2.   This square is made into a torus. See Figure 3.1

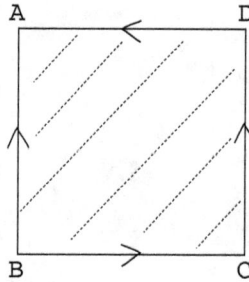

Figure 7.3.   This square is made into Klein bottle. See Figure 5.1

Attach the two segments, AB and DC, as drawn in Figure 7.4 so that the directions of the arrows coincide.

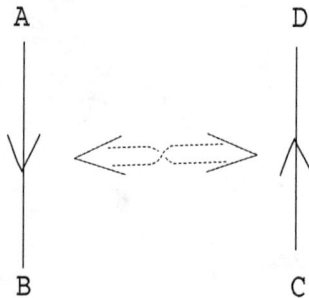

Figure 7.4.

Then the points A and C are identified and the points B and D are identified. Note that a 'half-twist' appears.

Attach two segments, BC and AD, as drawn in Figure 7.5 so that the directions of arrows meet.

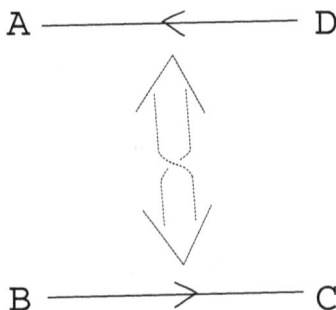

Figure 7.5.

Then the points B and D are identified and the points A and C are identified. Note that a 'half-twist' appears.

After these two operations, the four points A, B, C, and D all coincide.

Note that in this case a 'half-twist' appears again: When we make a Klein bottle, we get exactly one 'half-twist' from identifying opposite sides. However, in this case, we get two half-twists when we identify opposite sides.

You can bend and stretch the square $ABCD$ continuously without cutting or touching itself. We assume that the interior of the square does not touch itself. We suppose that the resulting object does not touch itself.

Question 7.1. Can you complete this construction? If possible, what do you obtain?

Attach two segments AB and DC as drawn in Figure 7.6 to obtain a Möbius band.

After that, can you attach the two edges DA and BC so that the directions of two arrows coincide? Or does that seem impossible?

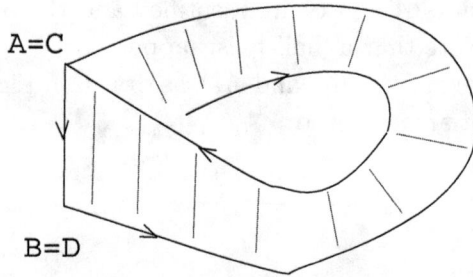

Figure 7.6.

How about this way instead? First, attach two segments, DA and BC, as drawn in Figure 7.7. We obtain a Möbius band again. After that, attach the two edges AB and DC so that the directions of two arrows agree.

This seems impossible, too.

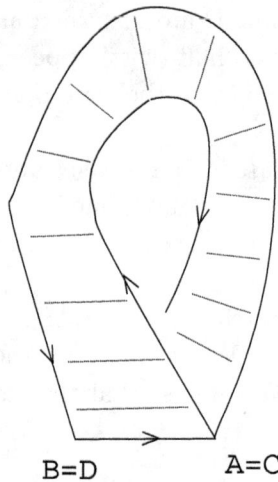

Figure 7.7.

Is there a method to do both operations (attaching AB to DC and attaching DA to BC) at the same time? This seems impossible, too.

Indeed, it is known that we cannot complete this construction of Question 7.1 on page 101 in three-dimensional space $\mathbb{R}^3$. Did you guess this was true?

Do you guess the following sentence, too?

It is also known that we can finish the construction of Question 7.1 in four-dimensional space $\mathbb{R}^4$.

Well, we obtain a figure in four dimensional space $\mathbb{R}^4$. We call it the *two-dimensional real projective space* $\mathbb{R}P^2$. Can you guess the shape of $\mathbb{R}P^2$? We revisit this later.

Recall the construction of the Klein bottle in Figure 7.8.

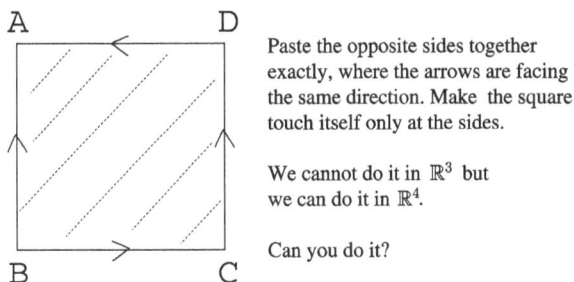

Paste the opposite sides together exactly, where the arrows are facing the same direction. Make the square touch itself only at the sides.

We cannot do it in $\mathbb{R}^3$ but we can do it in $\mathbb{R}^4$.

Can you do it?

Figure 7.8.   Question 5.1 on page 57. See also page 59

We cannot finish this construction in $\mathbb{R}^3$. However, if we allow the square to touch itself, we can complete it as in Figure 7.9.

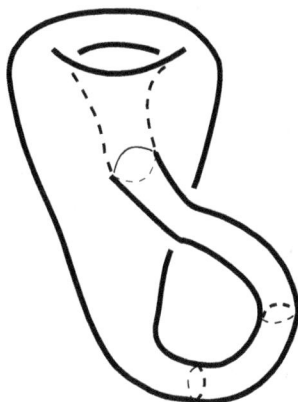

Figure 7.9.   The same figure as Figure 5.10

Here note that it intersects itself as drawn in Figure 7.10.

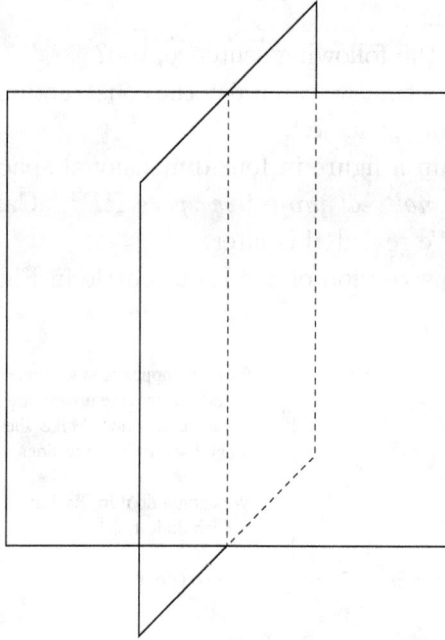

Figure 7.10.

Do you remember the DIY construction as in Figures 7.11 and 7.12?

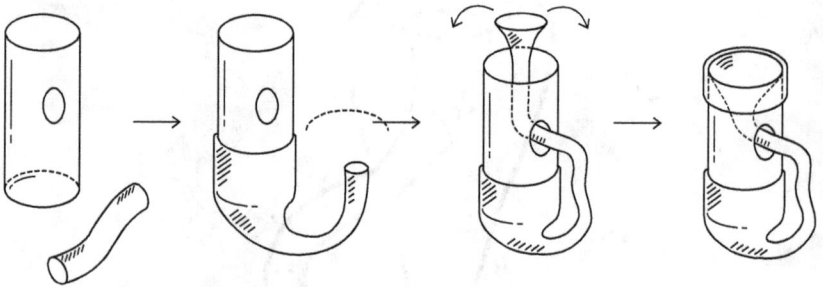

Figure 7.11.   The same figure as Figure 5.27

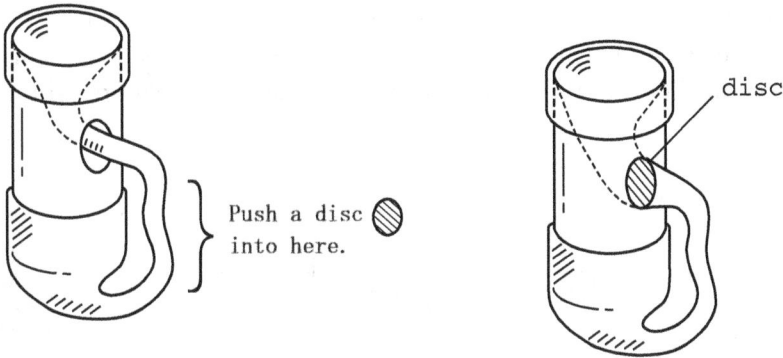

Push a disc into here.

disc

Figure 7.12.   The same figure as Figure 5.28

See Figure 7.10. At the place where the surface touches itself, two sheets of the surface intersect as drawn in Figure 7.10. This intersection is said to be made of *double points*.

We wrote earlier that we cannot construct $\mathbb{R}P^2$ in $\mathbb{R}^3$ without touching itself, associated with Question 7.1 on page 101 with comments on page 103; see Figure 7.13.

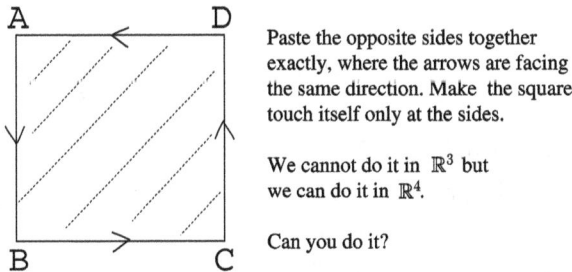

Paste the opposite sides together exactly, where the arrows are facing the same direction. Make the square touch itself only at the sides.

We cannot do it in $\mathbb{R}^3$ but we can do it in $\mathbb{R}^4$.

Can you do it?

Figure 7.13.   Question 7.1 on page 101. See also page 103

Now that we have the idea of double points, can we construct $\mathbb{R}P^2$ in $\mathbb{R}^3$ if we allow double point intersection?

What do you think?

It turns out that it is impossible.

Weakening the condition, consider the following problem.

In Figure 7.14, three sheets intersect at a *triple point.* These three sheets can be bent a little. If we allow not only double point curves shown in Figure 7.10 but also triple points shown in Figure 7.14, can we construct $\mathbb{R}P^2$ in $\mathbb{R}^3$?

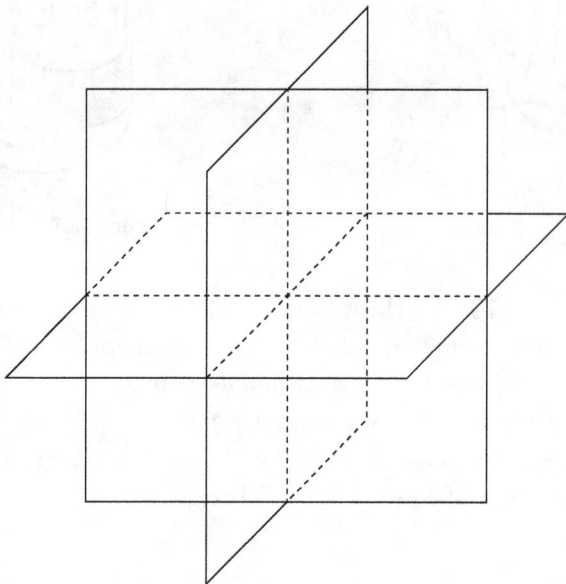

Figure 7.14.

What do you think?

Werner Boy discovered in 1901 that we can do it! His paper is cited in Further Reading on page 151. Its shape is drawn as in Figure 7.15. Boy is a great mathematician in Germany, who lived from May 4, 1879, until September 6, 1914. The figure is called *Boy surface* after his discovery. See Milnor and Stasheff's textbook in Further Reading for mathematical terms: $\mathbb{R}P^2$, immersions, etc. Page 82 of their book quotes Boy's paper. The author's introductory book includes Boy surface, which is the penultimate book in Further Reading.

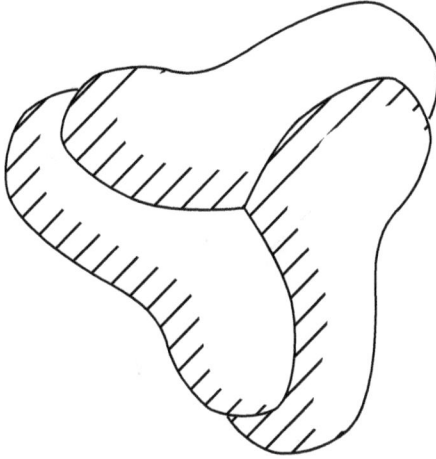

Figure 7.15. Boy surface

While Boy surface exists in $\mathbb{R}^3$, it is quite difficult to actually imagine this surface. You can find images of the surface form many viewpoints by searching the internet, which may help you gain an understanding of the shape. However, it is a little difficult for the beginners to imagine the shape of Boy surface because of its complicated shape.

Boy surface is as important as the Möbius band and the Klein bottle in mathematics. However, Boy surface is not well known as either the Möbius band or the Klein bottle.

One reason could be the complicated shape of Boy surface. The Möbius band can be made easily, as we did in §2.1. Many people understand the shape of the standard immersion of the Klein bottle in $\mathbb{R}^3$ although they have not contacted it directly. On the other hand, most people find it difficult or impossible to picture Boy surface in $\mathbb{R}^3$. If you make Boy surface on your own, you can understand the shape of Boy surface. We have created an easy way to construct Boy surface using a pair of scissors, a piece of paper, and a strip of scotch tape. We introduce the method in §7.4.

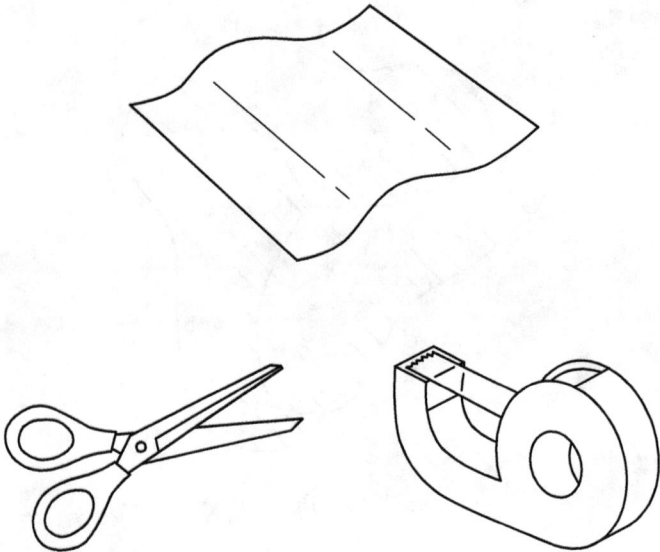

7.2.   The Two-Dimensional Real Projective Space $\mathbb{R}P^2$ and Four-Dimensional Space

Before making Boy surface in $\mathbb{R}^3$, we construct $\mathbb{R}P^2$ in $\mathbb{R}^4$. Recall a problem in Figure 7.16.

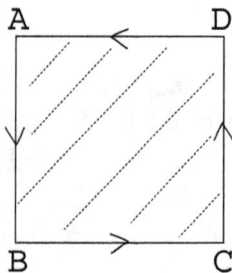

Paste the opposite sides together exactly, where the arrows are facing the same direction. Make the square touch itself only at the sides.

We cannot do it in $\mathbb{R}^3$ but we can do it in $\mathbb{R}^4$.

Can you do it?

Figure 7.16.   Question 7.1 on page 101. See also page 103

We go into four dimensional space again. Are you excited?

Imagine that in $\mathbb{R}^4$ the position is characterized by width, depth, height, and time.

As we did, make a square with arrows in Figure 7.17, into the Möbius band shown in Figure 7.18.

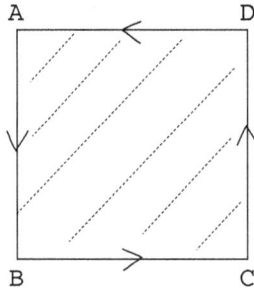

Figure 7.17.   The same figure as Figure 7.1

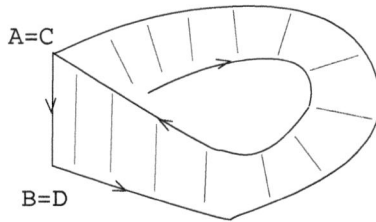

Figure 7.18.   The same figure as Figure 7.6

Put this Möbius band in $\mathbb{R}^3$, and then set time $t = 1$ to move into $\mathbb{R}^4$ as drawn in Figure 7.19.

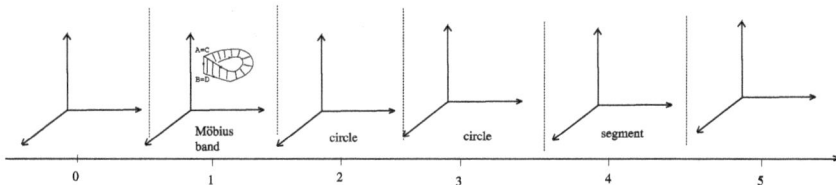

Figure 7.19.

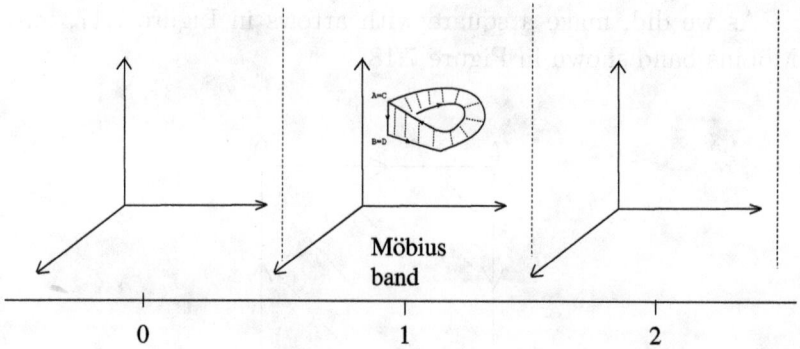

Figure 7.20. Figures 7.20 and 7.21 make one object

Flow the circle that is the boundary of the Möbius band at $t = 1$ in Figure 7.19, along time.

We enlarge Figure 7.19, and make a combination of Figures 7.20 and 7.21.

Flow the circle that is the boundary of the Möbius band at $t = 1$ in a combination of Figures 7.20 and 7.21, along time.

We omit both sides of a combination of Figures 7.20 and 7.21, and obtain Figure 7.22.

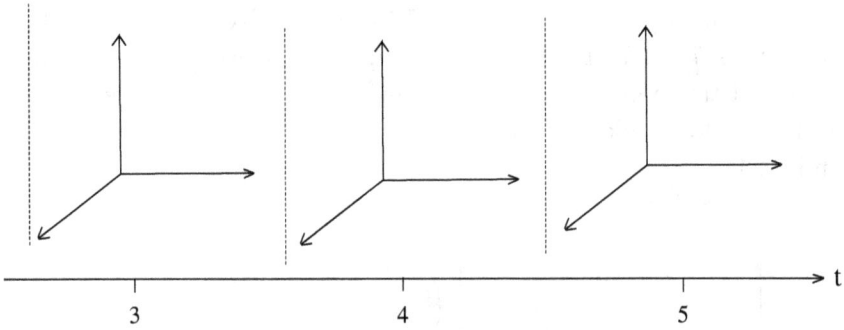

Figure 7.21. Figures 7.20 and 7.21 make one object

Flow the circle that is the boundary of the Möbius band in Figure 7.22, along time.

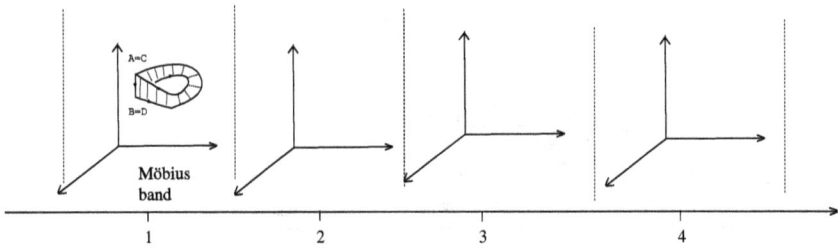

Figure 7.22.

We flow the circle in Figure 7.19 (respectively, a combination of Figures 7.20, 7.21, and 7.22) along time, however, in this case, it is a little more complicated than when we did this previously with the Hopf link and the Klein bottle, as shown in Figures 7.23 and 7.24.

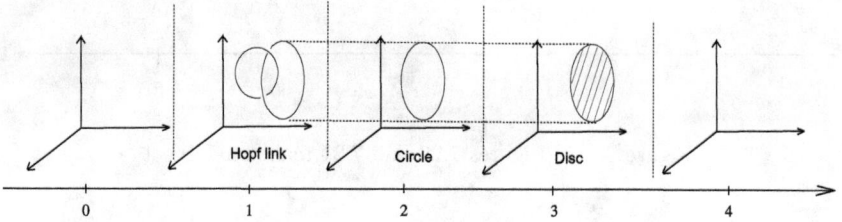

Figure 7.23.   The same figure as Figure 4.30

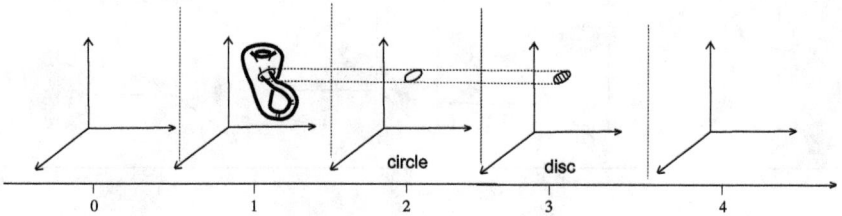

Figure 7.24.   The same figure as Figure 5.21

See a combination of Figures 7.25 and 7.26.

Next we move and put an object in four dimensional space $\mathbb{R}^4$. Are you ready?

Figure: A monster came from another dimension.

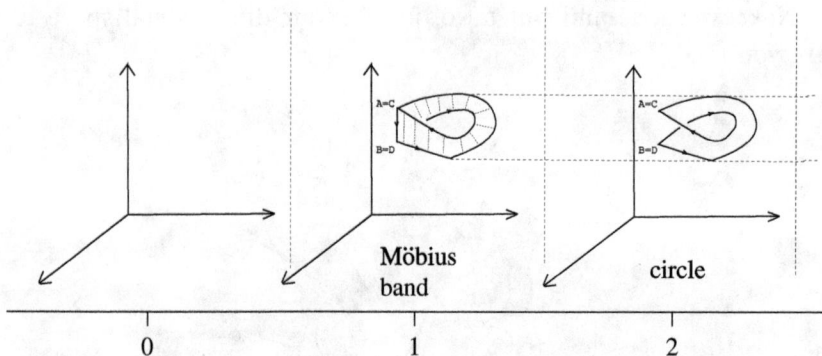

Figure 7.25.   Figures 7.25 and 7.26 make one object

We explain what happens in this figure, which is a combination of Figures 7.25 and 7.26 in the following.

Place the Möbius band at time $t = 1$. Move the boundary that is a circle along time until time $t = 2$. We do not transform the shape of the circle as we do in $2 < t$.

While moving the circle from $t = 2$ to $t = 3$, transform the circle continuously, a little by a little, without cutting or touching itself. At $t = 3$, we obtain the new shape drawn there.

Since time moves linearly in this construction, we can perform this transformation.

During the time interval from $t = 3$ to $t = 4$, the curve's segments, BC and DA, are deformed. At $t = 4$, two curved segments coincide. We did it!

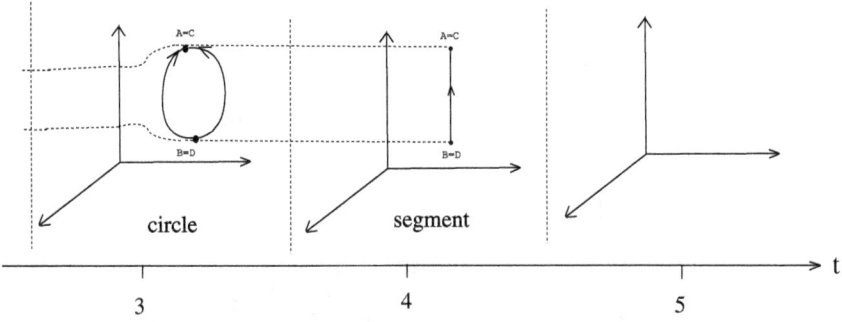

Figure 7.26. Figures 7.25 and 7.26 make one object

Indeed, the resulting object does not touch itself.

We have completed a construction of $\mathbb{R}P^2$ in $\mathbb{R}^4$ now.

We omit both ends of Figures 7.25 and 7.26, and draw Figure 7.27.

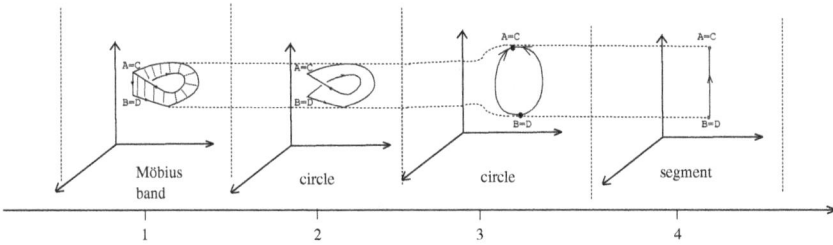

Figure 7.27.

Therefore, we have answered the question in Figure 7.28.

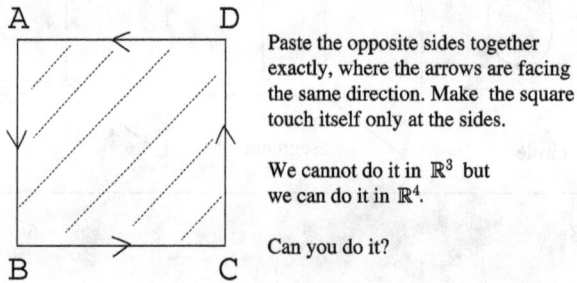

A   D

Paste the opposite sides together
exactly, where the arrows are facing
the same direction. Make the square
touch itself only at the sides.

We cannot do it in $\mathbb{R}^3$ but
we can do it in $\mathbb{R}^4$.

Can you do it?

B   C

Figure 7.28.   Question 7.1 on page 101. See also page 103

Does $\mathbb{R}P^2$ appear in your mind? Is your brain connected with $\mathbb{R}^4$?
Recall that Klein bottle is non-orientable (page 97). $\mathbb{R}P^2$ is also
non-orientable by the same reason.

7.3. Möbius Band and a Disc

Take a disc and a Möbius band, as shown in Figure 7.29.

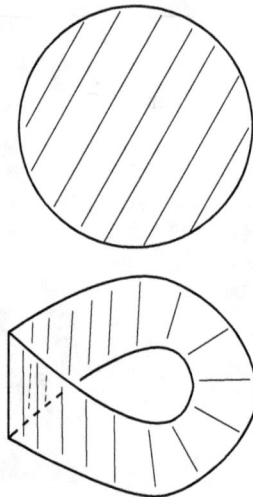

Figure 7.29.

The boundary of a disc is one circle. The boundary of a Möbius band is also only one circle. Recall Figure 6.15 in page 90.

Question 7.3. Can we glue these two boundary circles together with the following properties?

(1) The interiors of the disc do not touch itself, and the interior of the Möbius band does not, either.
(2) The interiors of the disc and the interior of the Möbius band do not touch each other.
(3) The resulting figure does not touch itself.
(4) You can bend and stretch both of them continuously without cutting itself.

If you think this is indeed possible, what object do you obtain?

The following question is much easier than Question 7.3.
Take two discs in Figure 7.30. The boundaries of both are circles.

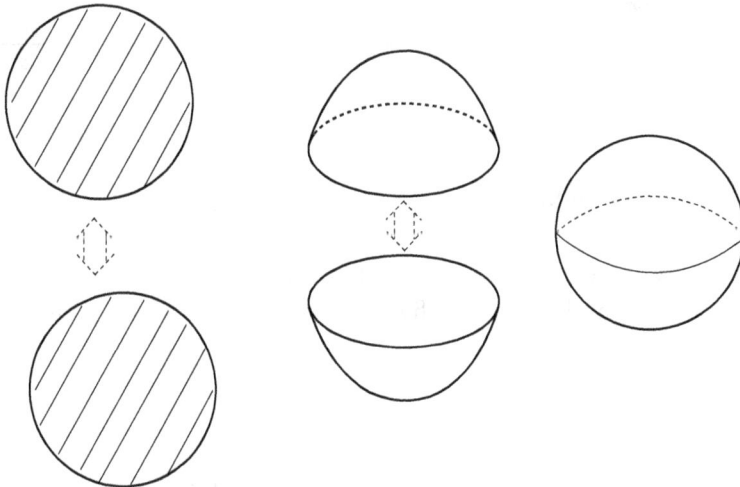

Figure 7.30.

Can we meet these two circle completely with the following properties?

(1) Neither of the two discs satisfies the condition that the interior touches itself.

(2) The interiors of the two discs do not touch each other.
(3) The resulting figure does not touch itself.
(4) You can bend and stretch both of them continuously without cutting itself.

If possible, what figure do you obtain?
This one is straightforward to visualize in $\mathbb{R}^3$. It is the sphere.
We now answer Question 7.3: You can do it in $\mathbb{R}^4$ as follows.
See Figure 7.32.
Take $\mathbb{R}^4$ where each point is determined by width, depth, height, and time.
Place the Möbius band at time $t = 1$.

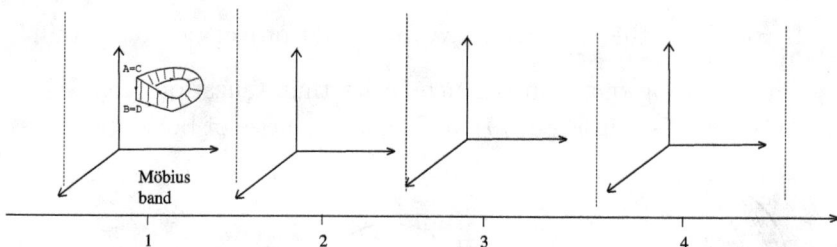

Figure 7.31.

Flow the boundary, which is a circle, from $t = 1$ to $t = 3$ as in Figure 7.32.
We only bend the Möbius band continuously without cutting or touching itself and make this figure, so this is the Möbius band, too.

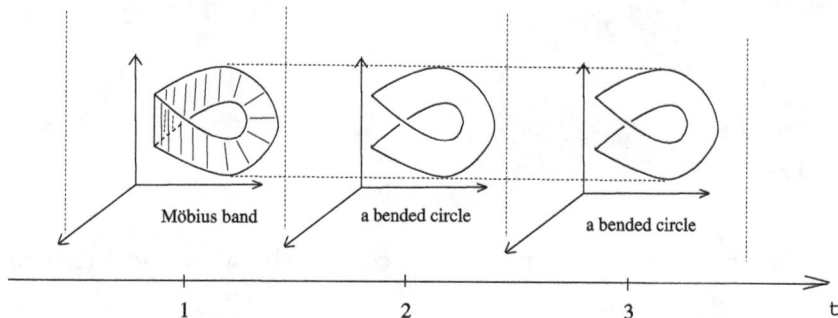

Figure 7.32.

Bend a disc and attach the bent disc at time $t = 3$ as in Figure 7.33.

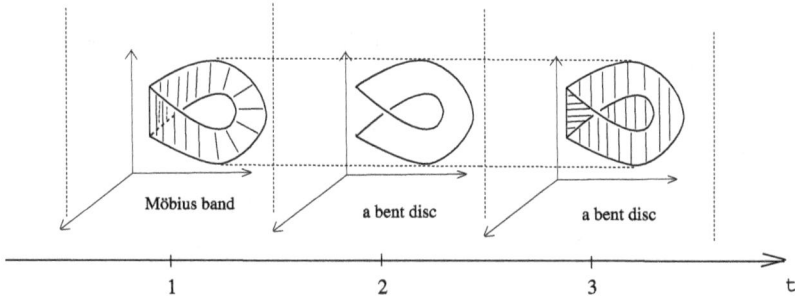

Figure 7.33.

Challenge to the reader: See that the object in Figure 7.33 is $\mathbb{R}P^2$ and that we stretch, bend, and move it without cutting or touching itself and obtain the object in Figure 7.27.

Since it is $\mathbb{R}P^2$, the construction in Question 7.3 on page 117 is impossible in $\mathbb{R}^3$.

7.4. Make Your Boy Surface

Boy surface is drawn as in Figure 7.34 as we stated in §7.1.

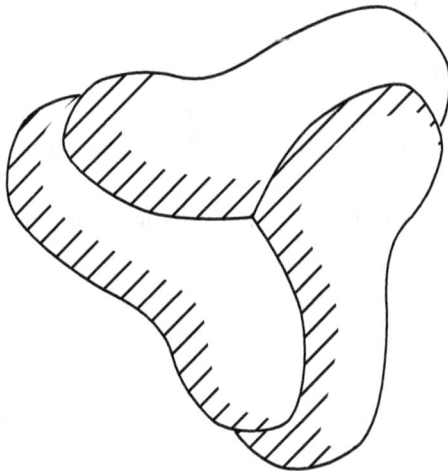

Figure 7.34. This figure is the same as one in Figure 7.15

We show a blueprint for Boy surface. Make your Boy surface!

We will use a pair of scissors, a piece of paper, and a strip of scotch tape to make Boy surface as we mentioned in §7.1. By constructing it on your own, you will gain a better understanding of Boy surface.

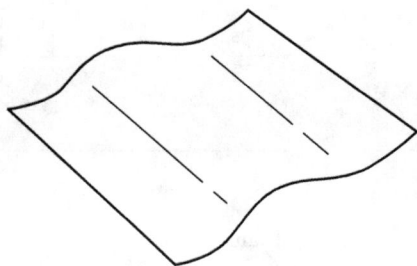

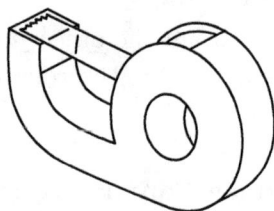

A blueprint of this paper construction is put in the following pages and on the internet. The readers can find these websites by typing in 'Eiji Ogasa', or 'Make your Boy surface'. The pdf is also available at https://arxiv.org/pdf/1303.6448.pdf

Download and print it, and do this paper construction on your own.

Furthermore, the author made a movie to demonstrate this paper construction. He put it in YouTube. You can also find it by typing in the author's name, 'Eiji Ogasa', or the title, 'Make your Boy surface'.

Let's begin.

See Figures I, II, and III. Make three copies of Figure I, a copy of Figure II, and three copies of Figure III.

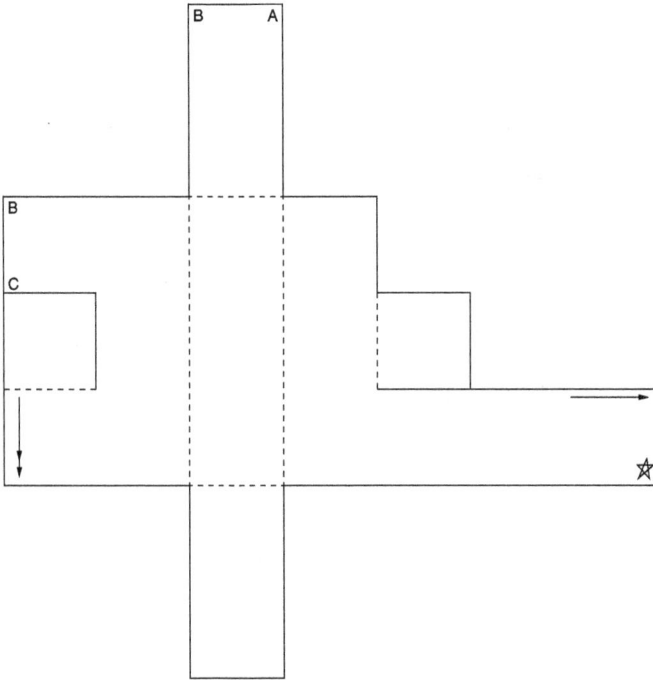

Figure I

Note: Make the copies so that the length of the edge of each of the unit squares in Figure I is half of that in Figures II and III. If it might be difficult to take such a copy of Figure II (resp. III), then we recommend the following way: Take a copy of Figure I at first. After that, make Figures II and III on a paper by using a scale and a pencil.

Figure III

Figure II

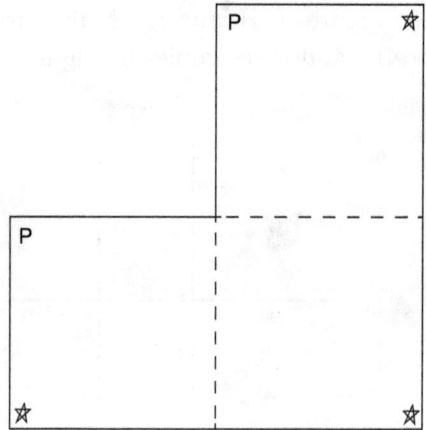

Make Figure IV from the three copies of Figure III. We call the result piece IV. Note that we must cut the three copies of Figure III a few times. If necessary, we can cut up one of the three copies of Figure III into multiple pieces and make piece IV from them by reattaching the pieces using scotch tape, being careful to remember exactly how they were cut so that you can reattach in the same exact manner.

If we imagine the x-, y-, and z-axes, then piece IV is the union of the following sets:

$$\{(x, y, z)| -1 \leq x \leq 1, \quad -1 \leq y \leq 1, \quad z = 0\}$$
and
$$\{(x, y, z)| -1 \leq y \leq 1, \quad -1 \leq z \leq 1, \quad x = 0\}$$
and
$$\{(x, y, z)| -1 \leq z \leq 1, \quad -1 \leq x \leq 1, \quad y = 0\}.$$

Take the following points in piece IV as shown in Figure V.

$$A = (-1, 0, 0), \quad B = (-1, 0, 1), \quad C = (0, 0, 1)$$
$$A' = (0, -1, 0), \quad B' = (1, -1, 0), \quad C' = (1, 0, 0)$$
$$A'' = (0, 0, -1), \quad B'' = (0, 1, -1), \quad C'' = (0, 1, 0)$$

We will use these points soon.

Piece IV

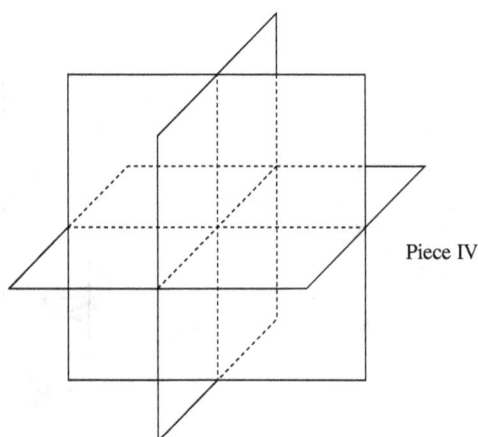

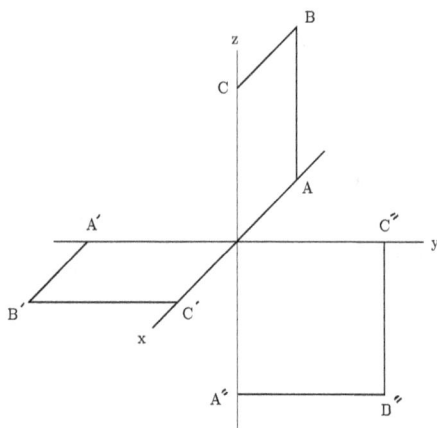

Cut each copy of Figure I along the solid lines. We will call the result piece I.

Fold piece I along the dotted line so that we see the dotted line inside and make "the angle made by the paper at the dotted line" exactly 90°.

Use a strip of scotch tape and attach the edges which meet. Note that the two points labeled *B* in Figure I meet. Then we obtain the following figure, which we refer to as piece I from now on.

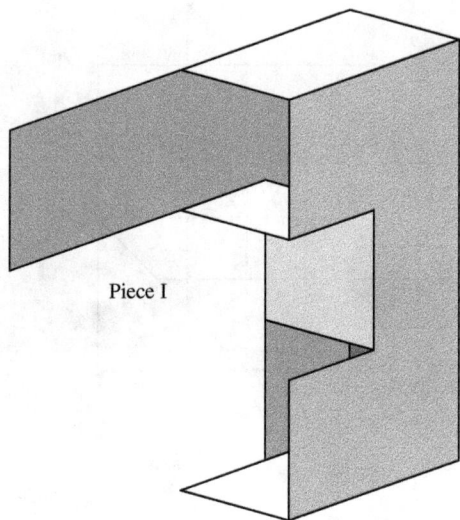

Piece I

Make three total copies of the piece I.

Call them the first piece I, the second piece I, and the third piece I.

Make sure that the points A, B, C are printed on each piece.

Attach the first piece I to piece IV with the following properties.

A meets A. B meets B. C meets C.

We obtain the following:

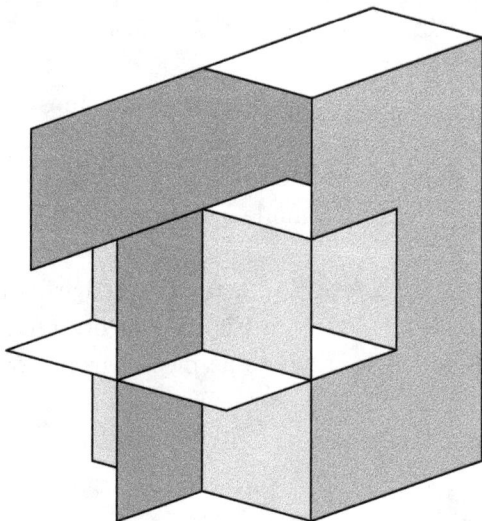

A, B, C in the second (resp. third) piece I are called A', B', C' (resp. A'', B'', C'').

Attach the second piece I to the piece IV (which already has the first copy of piece I attached) such that A meets A', B meets B', and C meets C'.

Attach the third piece I to the piece IV (which now has two copies of piece I attached) such that A'' meets A', B'' meets B', and C'' meets C'.

We obtain the following figure. It is called piece VI. Note that the arrow in a copy of the piece I meets one of the 'double arrows' in another copy of piece I.

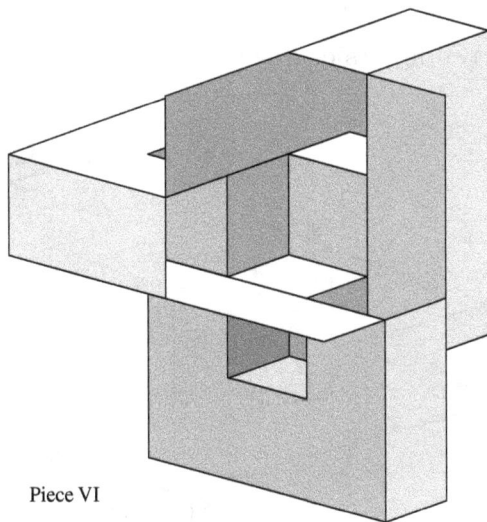

Piece VI

Cut a copy of Figure II along the solid lines. We will call the result piece II.

Fold piece II along the dotted line so that we see the dotted line inside and make "the angle made by the paper at the dotted line" 90°.

Use a strip of scotch tape and attach the edges which meet. Note that the two points labeled P meet. Then we obtain the following, which we will call piece II from now on.

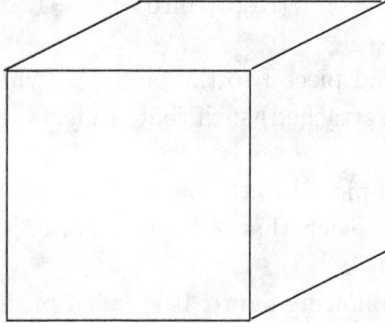

Piece II

Attach piece II to piece VI so that each star in piece II meets each star in piece VI. The result is Boy surface.

We include two figures of Boy surface, shown from two different perspectives.

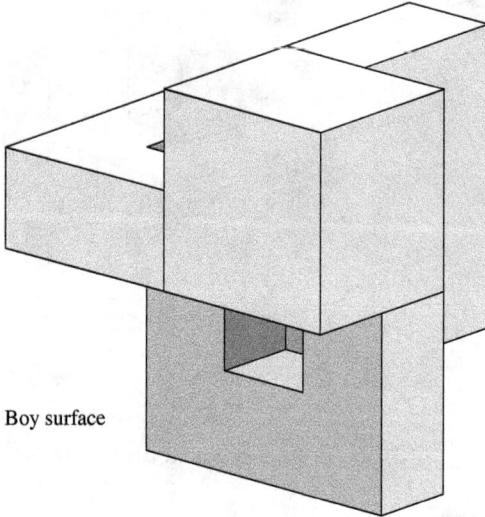

Boy surface

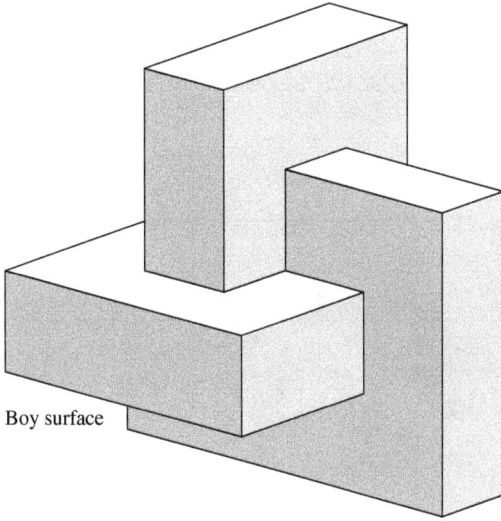

Boy surface

The line which is the intersection of two sheets in the following figure is the set of double points. (The double points are the intersection of two sheets.)

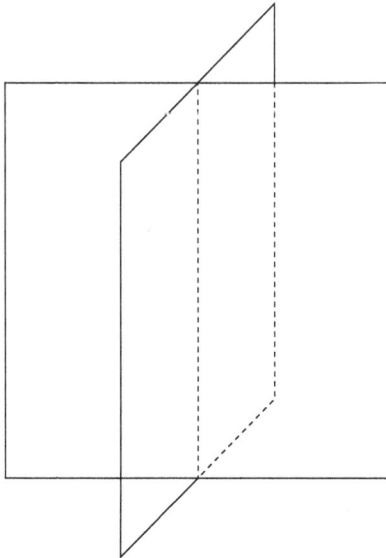

The point $(0,0,0)$ in piece IV is the triple point. (The triple point is the intersection of three sheets as shown in the following.) Boy surface contains only one triple point.

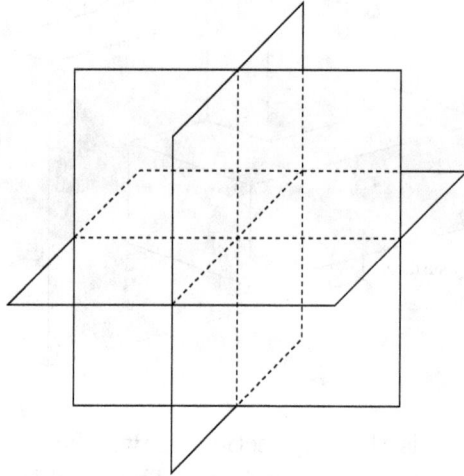

It is known that we cannot immerse $\mathbb{R}P^2$ into $\mathbb{R}^3$ without a triple point.

This Boy surface has a corner.

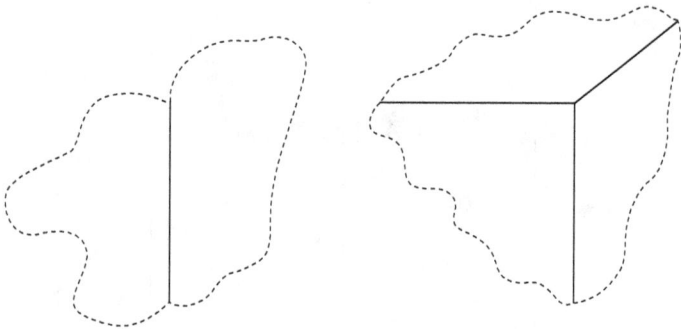

If you prefer Boy surface without corner, imagine making the corner smooth. Or, actually make it smooth in your paper construction. Then we obtain Boy surface shown in Figure 7.35.

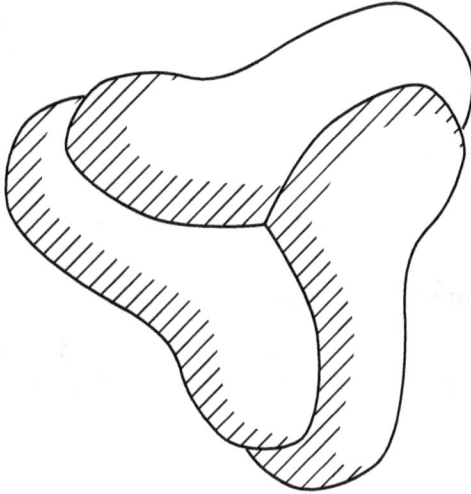

Figure 7.35. The same figure as Figure 7.15

7.5. Prove

We explain briefly how to prove that the paper construction which we made is indeed the immersion of $\mathbb{R}P^2$.

Proof Sketch: Remove piece II and piece IV from the paper construction of Boy surface. Then the result is made from the three copies of piece I. Remove the double point set by the following procedure.

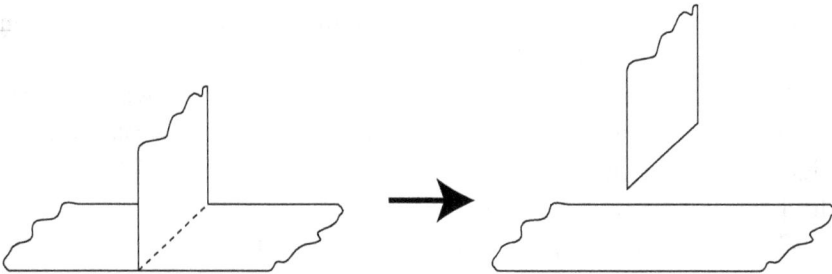

  Prove the result is (the Möbius band)−(three discs). Note that the boundary is a set of four circles. □

After the beginners learn homology groups and the Poincaré theorem on classifying surfaces by textbooks in Further Readings, they will understand the following.

Alternative Proof Sketch: Calculate the homology groups or the betti number of the manifold whose immersion is the Boy surface. After that, use the Poincaré theorem on classifying surfaces.   □

7.6.   Five-Dimensional Space

Recall how we embedded the Klein bottle in four-dimensional space $\mathbb{R}^4$. You may think of the four coordinates of $\mathbb{R}^4$ as width, depth, height, and time as we have done multiple times throughout this book. See Figure 7.36.

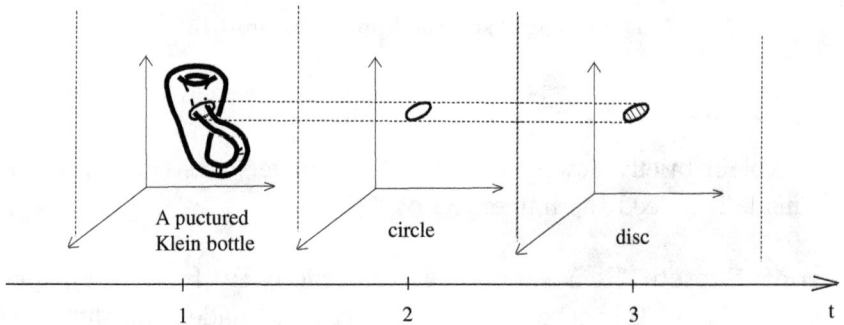

Figure 7.36.   The same figure as Figure 5.21

Put an immersion of Klein bottle (Figure 5.10) in the $t = 0$ section of $\mathbb{R}^4$. See Figure 7.37.

The following is known: In $\mathbb{R}^4$ you can move Klein bottle immersed as in Figure 7.36 to the immersed Klein bottle placed as in Figure 7.37. Furthermore, except for the last moment in Figure 7.37, the Klein bottle does not touch itself.

Can you make a procedure like this in the case of the two-dimensional projective space $\mathbb{R}P^2$?

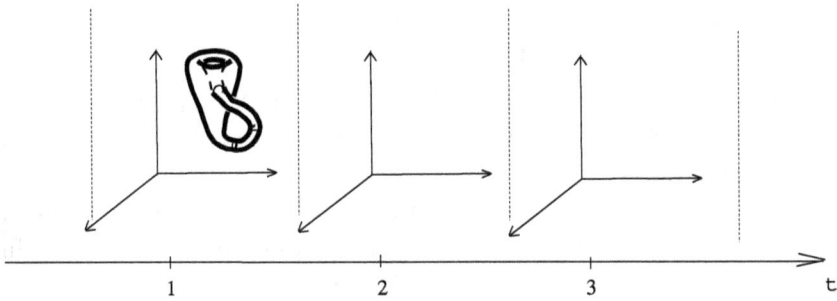

Figure 7.37. An immersion of Klein bottle (Figure 5.10) put in the $t = 0$ section of $\mathbb{R}^4$

Put $\mathbb{R}P^2$ in $\mathbb{R}^4$ as in Figure 7.38.

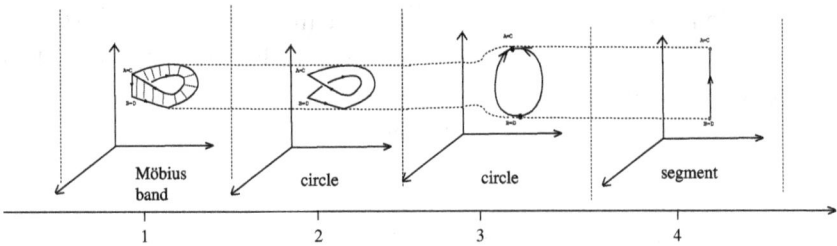

Figure 7.38. The same figure as Figure 7.27

You may put it as in Figure 7.33 because both $\mathbb{R}P^2$ are transformed into each other without touching itself in $\mathbb{R}^4$.

Place Boy surface in the $t = 0$ section in $\mathbb{R}^4$ as in Figure 7.39.

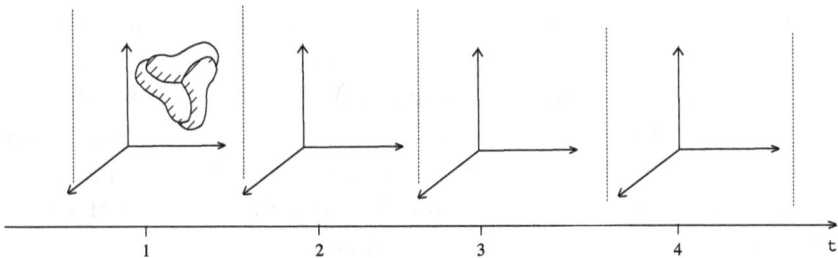

Figure 7.39. The object at $t = 0$ is the same one as Figure 7.15

Can you do the following procedure? In $\mathbb{R}^4$ you move $\mathbb{R}P^2$ put as in Figure 7.38 (respectively, Figure 7.33) to Boy surface, which is an immersed $\mathbb{R}P^2$, placed as in Figure 7.39. Furthermore, you must keep $\mathbb{R}P^2$ from touching itself until the very last moment in Figure 7.37. Is it possible to actually do this?

Could you guess the answer? It turns out that we cannot do it. (Strictly to say by using technical terms, no embedding of $\mathbb{R}P^2$ in $\mathbb{R}^4$ is regular homotopic to Boy surface in $\mathbb{R}^3$ of $\mathbb{R}^4$.)

However, we have the following. Take (or imagine) another direction different from width, depth, height, or time and construct *five-dimensional space* $\mathbb{R}^5$ by adding this extra direction. Regard $\mathbb{R}^4$ as a slice in this $\mathbb{R}^5$.

Then $\mathbb{R}P^2$ in Figure 7.36 and Boy surface in Figure 7.37 can also be regarded ones put in $\mathbb{R}^5$. In $\mathbb{R}^5$ you can move this $\mathbb{R}P^2$ to Boy surface so that except for the last moment in Figure 7.37, $\mathbb{R}P^2$ does not touch itself in any moment.

Can you see five-dimensional space $\mathbb{R}^5$?

Miscellaneous: High-dimensional Space Applied in Daily Life

Consider a stick-shaped cookie. Assume that our cookie is so thin that we may regard it mathematically as a line segment. Measure the temperature at each point of the cookie and draw a graph which associates each point of the cookie with its measured temperature. The graph is shown in Figure 7.40. Note that our graph lives in two-dimensional space.

Next, consider a board-shaped chocolate, where this time we will assume that our chocolate is so thin that we may regard it mathematically as a filled-in rectangle. Measure the temperature at each point of the sheet of chocolate, and draw the corresponding graph which associates to each point of the rectangle the measured temperature. This graph is shown in Figure 7.41. Note that this time our graph lives in three-dimensional space.

Now, take our dessert to be a solid rectangular cake. We again measure the temperature at each point of our dessert and draw

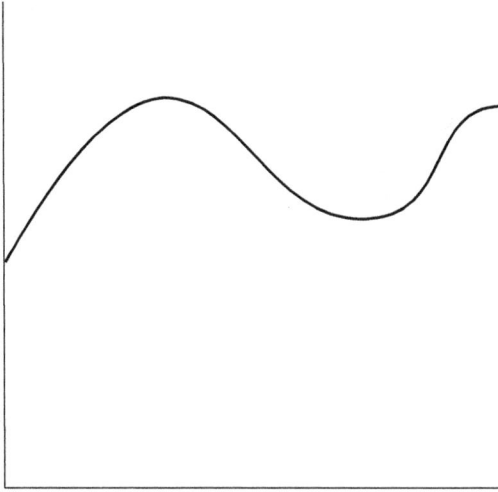

Figure 7.40.

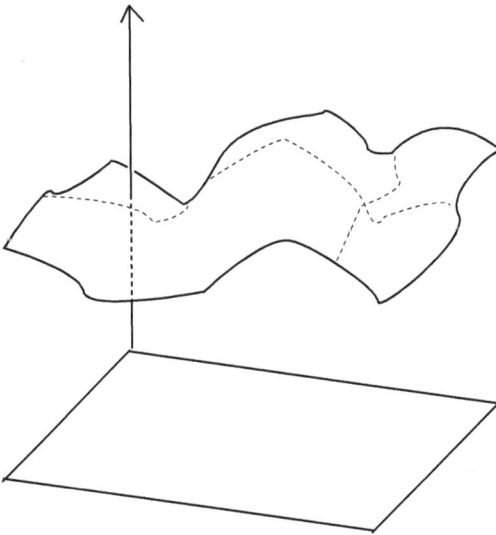

Figure 7.41.

the corresponding graph with each point of the cake associated to the measured temperature. This time, where does our graph live?

Following the pattern, it becomes clear that our graph lives in four-dimensional space. Try to visualize the graph in your mind.

Define now the *sugar density* at a point A to be how many grams of sugar there are in the cubic centimeter centered on A. Now, suppose we assign to each point of the above cake both the temperature and the sugar density. In this case, where does our graph live? You may guess that along the same lines, adding another bit of information about our dessert takes our graph up one more dimension. The answer is that this graph lives in five-dimensional space. Now try to visualize this graph in your mind.

In our daily lives, we make use of correspondence (that is, functions and maps) from many variables to many values. For example:

$$f(x, y) = x^3 y^2 \sin xy$$

$$(a(x, y, z), b(x, y, z)) = \left(y^x + \log z, xy + \frac{z^2}{y^3} \right).$$

We need these maps when we calculate our taxes, insurance costs, or interest, or perhaps when we consider the theory underlying the functionality of our computers. If we want to draw the graphs of those maps, we have to imagine them in high-dimensional spaces.

Moreover, considering maps from many variables to many values is equivalent to considering graphs in high-dimensional spaces. In many cases, we can study objects in high-dimensional spaces to get information about the properties of certain multi-variable, multi-valued maps. It is thus necessary to have an understanding of how to apply high-dimensional mathematics to our daily lives. In order to do this, people typically study *manifolds*, which we introduce in the following section, Chapter 8.

Chapter 8

Manifolds

We now answer a question posed in the Introduction: There is a possibility that our universe has the following property. If you leave Earth in a space ship and travel in one direction away from Earth, you will ever end up back at Earth again. Figure 8.1 is a sci-fi depiction of this scenario. Under this assumption, what is the shape of our universe?

Figure 8.1. The same figure as Figure 1 on page ix.

8.1. Two-Dimensional Manifolds

If you zoom in on a small part of the sphere, it looks just like a plane (with just a little bit of curvature, but we can ignore that for now).

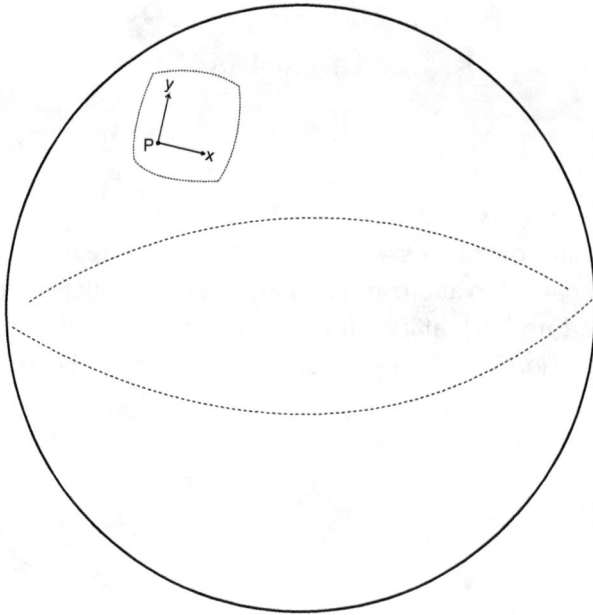

Figure 8.2.   The sphere

Assume that a figure X has the following property(∗). Note that the sphere in Figure 8.2 has this property.

(∗) For any point P in X, there is a small neighborhood around P that looks like a small plane $\mathbb{R}^2$, possibly curved a little bit. Note that the neighborhood includes P inside and that P is a point of the neighborhood.

Then we call X a 2-dimensional manifold. Note that we can draw the x- and y-axes on U as in Figure 8.2, and that we can regard U as a 'small' $\mathbb{R}^2$.

A plane $\mathbb{R}^2$, the sphere, the Klein bottle, and 2-dimensional real projective space $\mathbb{R}P^2$ are all 2-manifolds. In fact, there are infinitely many 2-manifolds!

The interior of the Möbius band is also a 2-manifold. Note that when we consider Möbius band with the boundary, it is an example of 2-dimensional *manifold with boundary*.

We define *1-(respectively, 0-) dimensional manifolds* by using $\mathbb{R}^1$ (respectively, $\mathbb{R}^0$) in the same fashion. Note that $\mathbb{R}^0$ is a point.

See the textbooks in Further Reading on page 151 for more details about manifolds (with boundary).

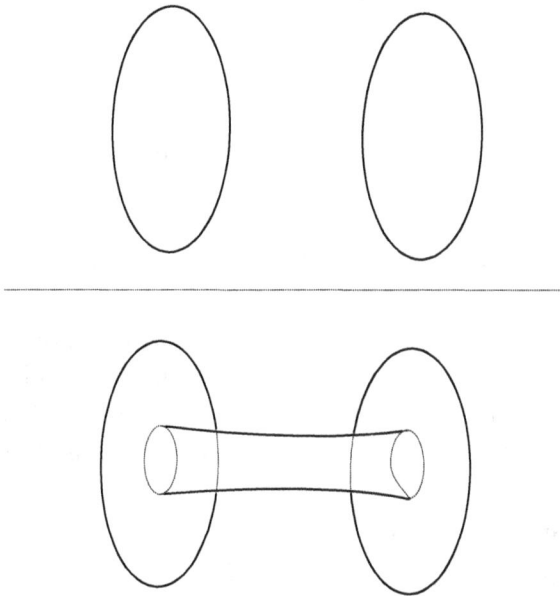

Figure 8.3. The connected sum

See Figure 8.3. Take two 2-manifolds, and note a disc that is a part of each 2-manifold. Change the two discs as the upper object of Figure 8.3 is made into the lower of there. Then we obtain one 2-manifold from the two 2-manifolds. This operation is called a *connected-sum*.

The connected-sum of two torus is the double torus, which is the right object of Figure 8.4.

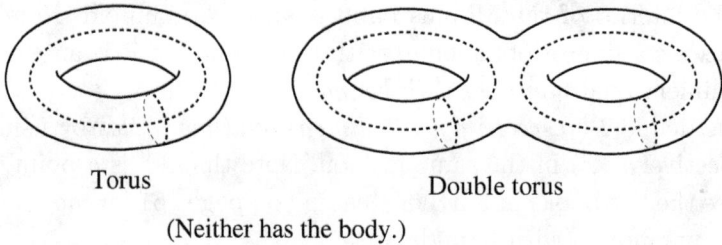

Torus   Double torus

(Neither has the body.)

Figure 8.4. The torus and the double torus

What is the connected-sum of two 2-dimensional projective space $\mathbb{R}P^2$?

The answer is the Klein bottle. Indeed we have explained already. Define an open disc to be a disc that does not have the boundary and have only the body. By Section 7.3, $\mathbb{R}P^2 - D^2 =$the Möbius band, where D^2 is an open disc. By Section 6.5, Möbius band + Möbius band= the Klein bottle.

Poincare proved that all closed 2-manifolds are made from ν copies of tori and μ copies of $\mathbb{R}P^2$ by a finite sequence of connected-sums, where μ and ν are non-negative integers. The word 'closed' is a mathematical term. It is Poincare's theorem on classifying surfaces.

8.2. The 3-Dimensional Sphere S^3

In this section, we give one answer to the problem proposed in the Introduction.

Under the condition of the problem, the universe has the following property.

($\sharp$) At any point P in the universe, there is a small neighborhood U around P in the universe that is a small $\mathbb{R}^3$. Note that U includes P and that P is a point of U. However, the universe is not $\mathbb{R}^3$.

There is a possibility as drawn in Figure 8.5.

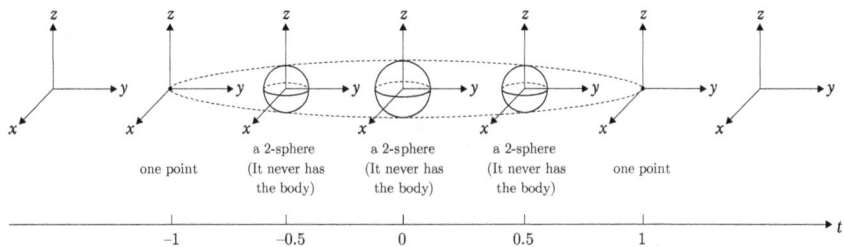

Figure 8.5. A candidate for the shape of the Universe drawn in four dimensional space $\mathbb{R}^4$

Can you imagine this figure in four-dimensional space $\mathbb{R}^4$? This is a one-dimensional higher analogue of the 2-dimensional sphere. Envision the figure in four-dimensional space of Figure 8.5 by seeing the 2-sphere in three-dimensional space of Figure 8.6.

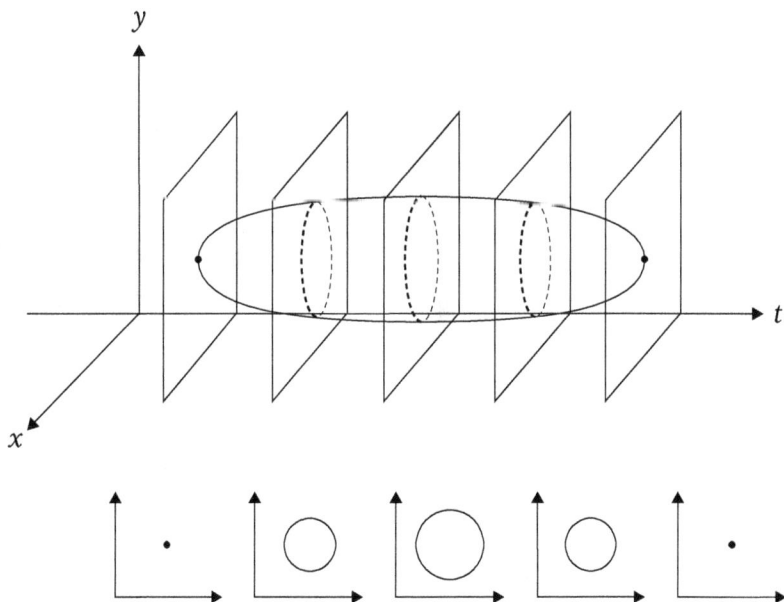

Figure 8.6. The 2-sphere in $\mathbb{R}^3$ and its sections

The 2-sphere is made from two curved discs, which we can see in Figures 8.7 and 8.8.

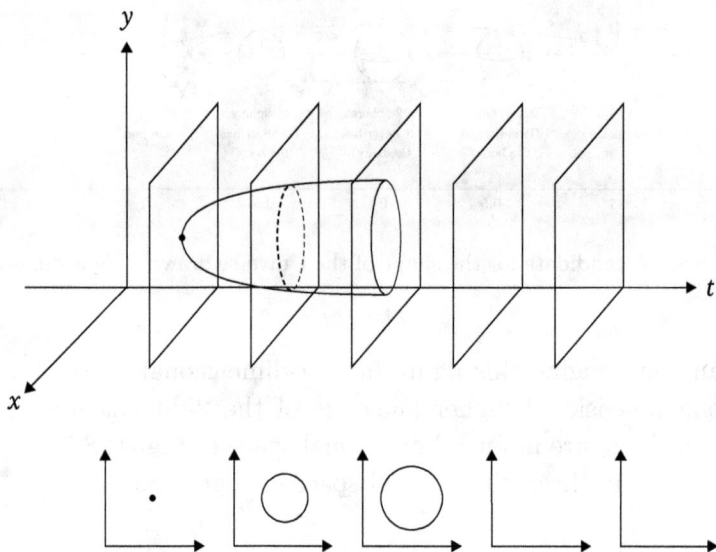

Figure 8.7.   A left semi-sphere of the sphere in Figure 8.6

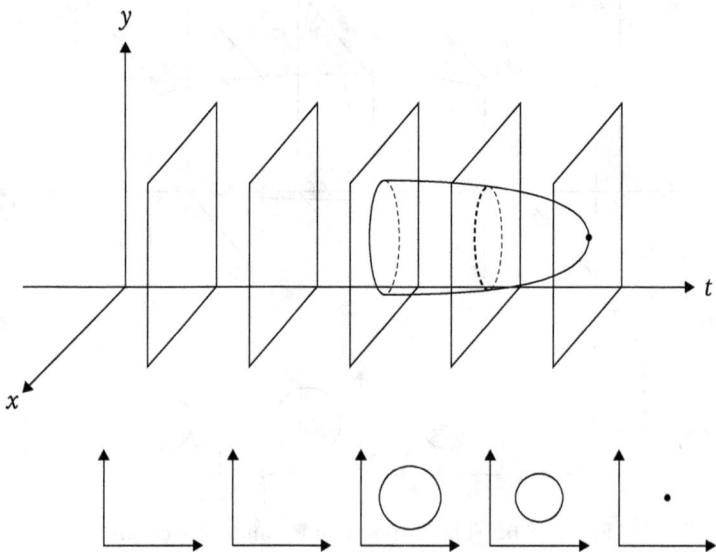

Figure 8.8.   A right semi-sphere of the sphere in Figure 8.6

We can increase the dimension by one as follows. The object in four-dimensional space of Figure 8.5 is composed of two objects: one in Figure 8.9 and the other in Figure 8.10. Indeed, we transform each of the two objects without cutting it or touching itself, and make it into Figure 8.11. Thus, it is the ball as the semi-sphere is transformed into a disc. Can you see it?

Figure 8.9. The left half of the object in Figure 8.5

Figure 8.10. The right half of the object in Figure 8.5

Figure 8.11. A ball in three dimensional space $\mathbb{R}^3$: It has the body and the boundary.

In fact, the object in Figure 8.5 satisfies the condition (♯) on page 138. Can you see why? We explain in the following. See Figures 8.12 and 8.13.

In Figures 8.12, we put the sphere and the two discs from which the 2-sphere is made, that is, the sphere = (the disc)+(the disc). We take a few points from the sphere and draw a disc around each point. You can think that a point moves in the 2-sphere. We draw them in both of the sphere and a set of the two discs.

sphere = disc + disc
(It never has the body.)

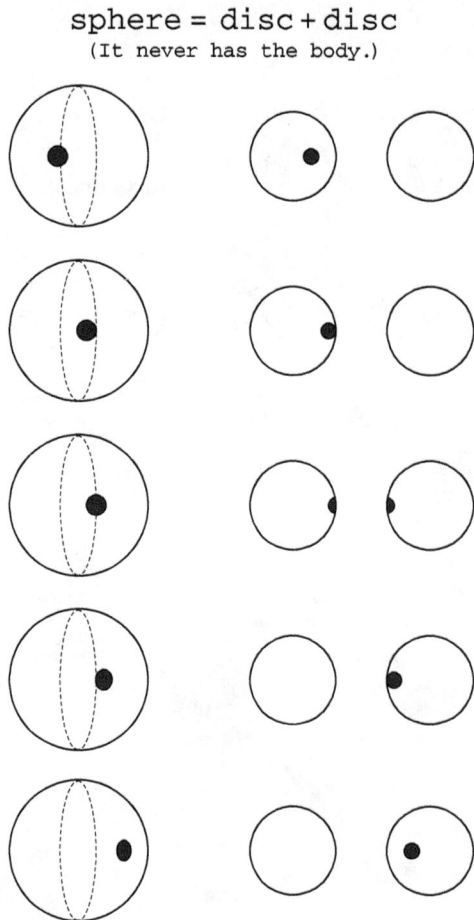

Figure 8.12.   The sphere = (the disc) + (the disc)

One-dimensional higher analogue of Figure 8.12 is Figure 8.13. In
Figures 8.13, we put the two balls from which the object is made.
They implicitly resent the object in Figure 8.5. That is, the object
= (the ball) + (the ball). We take a few points from the object and
draw a ball around each point. We draw them in a set of the two
balls. Imagine the balls around the points in the object from Figures
8.13. You can think that a point moves in the object.

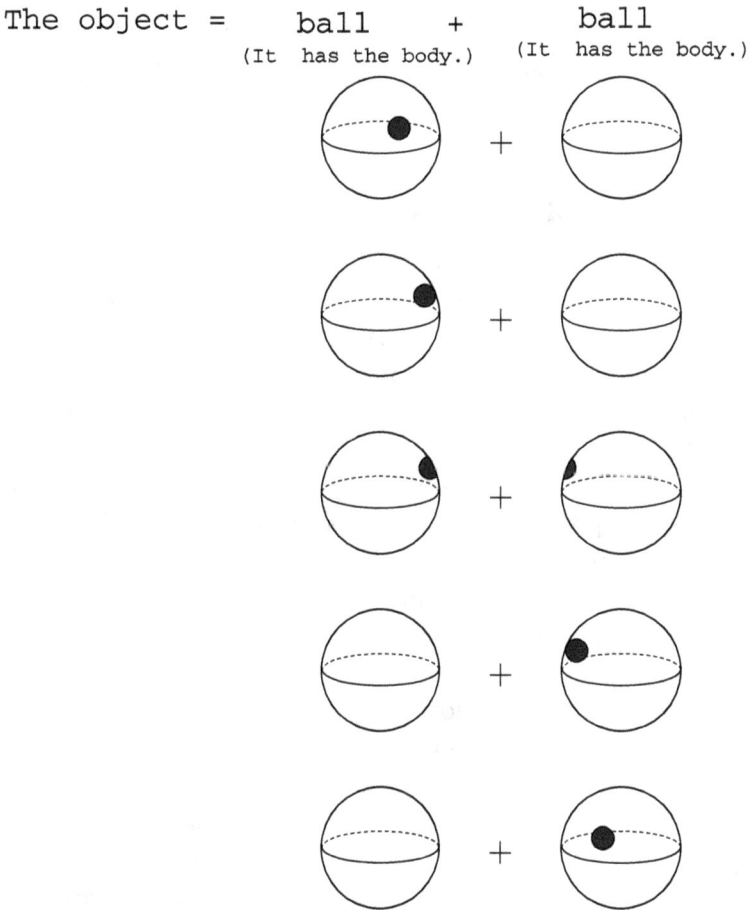

The object = ball + ball
(It has the body.) (It has the body.)

Figure 8.13. The object in Figure 8.5 = (the ball)+(the ball)

The mathematical object in Figure 8.5 is called the *3-dimensional sphere* S^3 or the *3-sphere*. We do not consider whether the 3-sphere is in $\mathbb{R}^4$ or not. We do not take care where the 3-sphere is, and instead only consider the 3-sphere itself. See textbooks on manifolds and topology for more details.

Then what we call the sphere is called the *2-dimensional sphere*, or the *2-sphere* S^2. Of course, we made the technical term, the 3-sphere, because the 3-sphere is a generalization of the 2-sphere.

We said that the shape of universe may be the 3-sphere, but is there another possibility? We will give more possible examples after a brief aside. Don't skip the following Miscellaneous. It is a hint of this question.

Miscellaneous: Black Holes, White Holes, and Worm Holes

Using Einstein's theory of general relativity, Schwarzschild predicted the existence of an astronomical object that has such intense gravity that it sucks in everything around it. After his prediction, astronomers actually observed such an object! Today we call it a *black hole*.

You might wonder where things go after they drop into a black hole. A potential sci-fi answer is as follows: A black hole is connected with another astronomical object, a *white hole*, by a tube, which we call a *wormhole*. See Figure 8.14. Things that are sucked into a black hole go through a wormhole and appear out of a white hole.

Neither a white hole nor a wormhole has been observed yet. Sometimes we consider a mathematical model even if it is almost impossible for such a model to actually exist. However, when we consider such a model, we might obtain new ideas in mathematics and physics. We call such a model a *toy model*.

Sometimes a toy model is realized in a different place or in a different form. In many other cases, even if it is not realized, it promotes the development of mathematics and physics. A wormhole

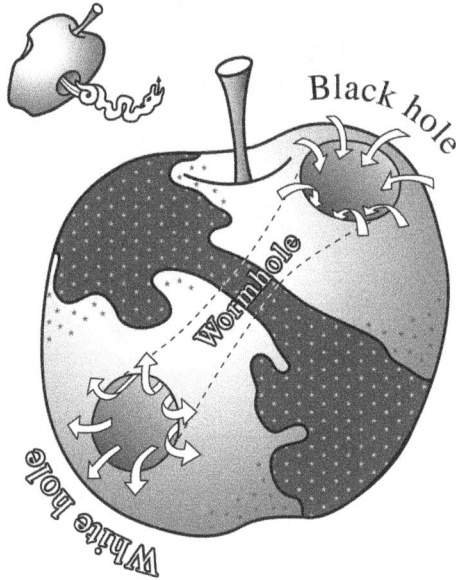

Figure 8.14. A black hole, a white hole, a wormhole and our universe

is an example of such a toy model, but it appears in current papers and helps drive new research in mathematical physics[i].

8.3. $S^1 \times S^2$

We answer the question on page 144 in the following. In Figure 8.14, we explain the case where we regard the universe as the 3-sphere with some mathematical changes. However, we draw a universe as the 2-sphere for simplicity. Can we draw this universe as a 3-dimensional

---

[i]For example, see the recent papers:

I. Bah, Y. Chen, and J. Maldacena: Estimating global charge violating amplitudes from wormholes, *Journal of High Energy Physics*, 61 (2023).

M. S. Morris and K. S. Thorne: Wormholes in spacetime and their use for interstellar travel: A tool for teaching General Relativity, *American Journal of Physics* 56, 395 (1988).

M. S. Morris, K. S. Thorne, and U. Yurtsever: Wormholes, time machines and the weak energy condition *Physical Review Letters* 61, 1446 (1988).

one, as we can draw the 3-sphere, the Klein bottle, $\mathbb{R}P^2$ although we cannot draw them in $\mathbb{R}^3$? Yes, we can.

In Figure 8.14, the 2-sphere S^2 becomes the torus when we include a black hole, a white hole, and a wormhole. See some sections of the torus drawn in Figure 8.15. We generalize Figure 8.15 and draw a universe with a black hole, a white hole, and a wormhole.

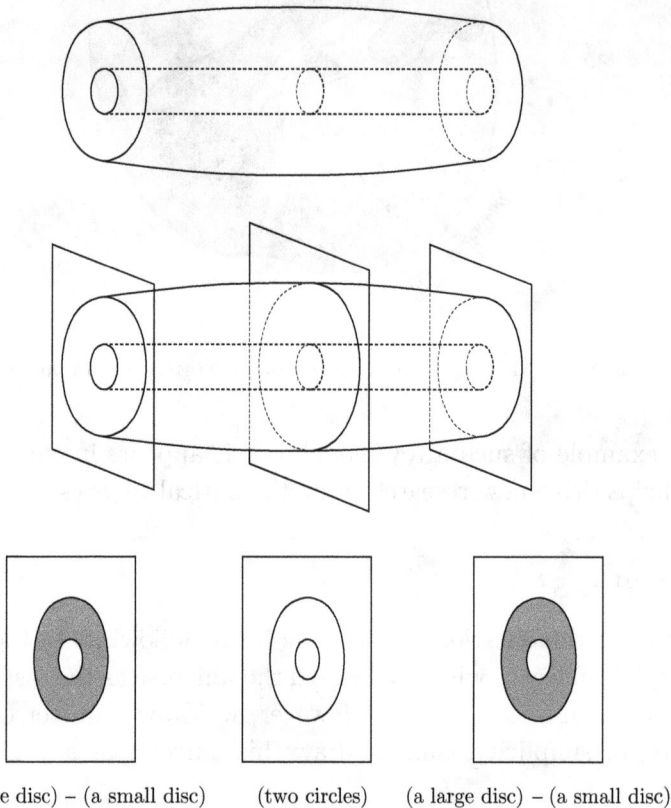

(a large disc) − (a small disc)   (two circles)   (a large disc) − (a small disc)

Figure 8.15.   The torus and its sections

Figure 8.16 is a one-dimensional higher analogue of Figure 8.15. The upper object in Figure 8.16 represents the 3-sphere. The lower object in Figure 8.16 is a universe with a black hole, a white hole, and a wormhole. This figure is called *the product manifold $S^1 \times S^2$ of* the circle S^1 and the (2-dimensional) sphere S^2, or simply $S^1 \times S^2$. One way of saying, $S^1 \times S^2$ is a locus that S^2 moves along S^1 in $\mathbb{R}^4$. After the beginners learn product manifolds, they will understand the meaning of $\times$ in $S^1 \times S^2$.

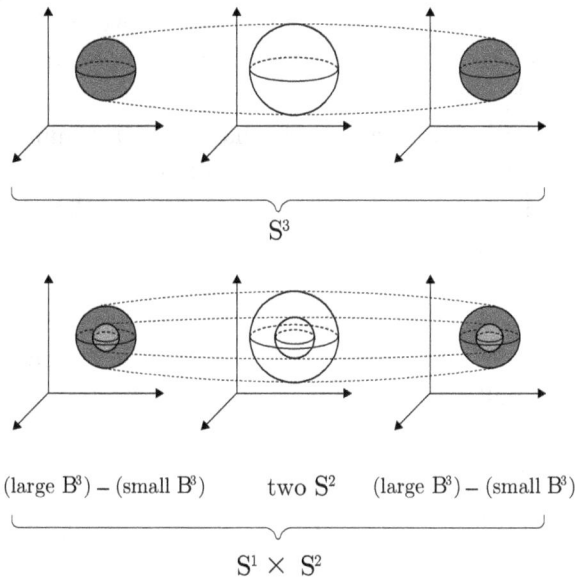

S^3

(large B³) – (small B³)   two S^2 (large B³) – (small B³)

$S^1 \times S^2$

Figure 8.16. S^3 and $S^1 \times S^2$

8.4. Three-Dimensional Manifolds

If a mathematical object M satisfies the condition that every point P has a neighborhood U around P that is a small $\mathbb{R}^3$, then M is called a *3-dimensional manifold*, or simply a *3-manifold*. Note that U includes the point P inside and that P is a point of U. Three dimensional space $\mathbb{R}^3$, the 3-sphere S^3, and $S^1 \times S^2$ are 3-dimensional manifolds. There are infinitely many 3-manifolds. Another example is the 3-dimensional torus $S^1 \times S^1 \times S^1$ or T^3. (The torus is also called the 2-dimensional torus $S^1 \times S^1$ or T^2.)

There is a possibility that the universe is not $\mathbb{R}^3$ or S^3 and instead is some other 3-manifold.

The sphere and the torus can be embedded in $\mathbb{R}^3$ as we see. (Note that the term 'embed' is a mathematical term, and the interested reader can look up this term online.) Although the Klein bottle and Boy surface cannot be embedded in $\mathbb{R}^3$, we can embed them in $\mathbb{R}^3$ if we remove a disc from them. In general, it is known that we can embed any 2-manifold X in $\mathbb{R}^3$ after we remove a disc from X, if necessary. Note that the dimension of $\mathbb{R}^3$ is larger than the dimension of 2-manifolds by one.

On the other hand, it is known that not all 3-manifolds can be embedded in $\mathbb{R}^4$, even if we remove a 3-ball from the 3-manifolds. (We cannot do this even if we require that the 3-manifold is 'orientable'!) Such an example is the *3-dimensional real projective space* $\mathbb{R}P^3$.

The *Poincaré 3-sphere* is a 3-manifold. The Poincaré 3-sphere can be embedded in $\mathbb{R}^5$ and cannot in $\mathbb{R}^4$ but can be embedded in $\mathbb{R}^4$ if we remove a 3-ball from it.

Readers who are interested in going into a higher level than this book can start by searching the names of the above manifolds in the internet.

8.5. High-Dimensional Manifolds

For each integer $n > 4$, $\mathbb{R}^n$ is defined as a generalization of $\mathbb{R}^m (m = 0, 1, 2, 3, 4)$. If a mathematical object M satisfies the condition that every point P has a neighborhood U around P that is a small $\mathbb{R}^n$, where n is a natural number, then M is called an *n-dimensional manifold*, or simply an *n-manifold*. Note that U includes P and that P is a point of U.

We call n-manifolds *high-dimensional manifolds* if n is larger than three or four. It depends on the situation which convention is used, $n \geq 4$ or $n \geq 5$.

There are still many open problems about n-manifolds. Some important ones are as follows:

The four-dimensional smooth Poincaré conjecture.

Some parts of high-dimensional smooth Poincaré conjecture.

The sliceness problem of even-dimensional links.

The μ invariant problem of 2-links, and problems that involve a natural higher-dimensional generalization of the 2-link case (see the author's paper Ogasa, 2004).

A 1-handle problem on smooth simply connected 4-manifolds (See Kirby's book in Further Reading).

Is there a smooth 4-manifold on which there are only finite smooth structures?

The ribbon-slice problem on 1-knots in the smooth category

And many, many more.

You can begin to learn more about these problems by searching the terms on the internet. Go forth and learn how to attack these problems. Solve them! Be a hero!

Miscellaneous: String Theory

String theory is a combination of the theory of relativity (both special relativity and general relativity) and quantum mechanics (and its generalizations, quantum field theory, and supersymmetric quantum field theory). String theorists believe that Universe and Nature are explained by string theory.

According to string theory, the universe is 10-(or 11-)dimensional space, or in mathematical terms, a 10-(or 11-)manifold. One possibility is that local part of the universe is $\mathbb{R}^{10}$: A very small copy of $\mathbb{R}^6$ occurs at each point of $\mathbb{R}^4$. Another possibility is that very small six-dimensional manifolds (possibly more complicated than 6-manifolds) occur each point of $\mathbb{R}^4$. Human beings experience only $\mathbb{R}^4$ in daily life because 6-dimensional part is too small for people to notice in daily lives. See textbooks on string theory for more detail.

One possibility for a 6-manifold that could be used as above is the Calabi-Yau 3-fold. This space is deeply connected with complex manifold theory and symplectic geometry, as well as knot theory and gauge theory in low dimensional topology.

This is one reason why we need to research high-dimensional manifolds. You have also embarked on the study of manifolds in all dimensions—now keep learning more!

Movies

We made movies of the paper constructions in this book and posted them on YouTube. You can find them by searching the author's name, Eiji Ogasa. Watch them!

Further Reading

J. W. Milnor and J. D. Stasheff: *Characteristic Classes*, Annals of Mathematics Studies, Princeton University Press, USA 1974.

J. W. Milnor: *Morse Theory*, Annals of Mathematics Studies, Princeton University Press, USA 1963.

R. C. Kirby: *The Topology of 4-Manifolds*, Springer, German 1989.

L. H. Kauffman: *Knots and Physics*, Knots and Everything, World Scientific Pub Co Inc., Singapore 1991.

L. H. Kauffman: *On Knots*, Annals of Mathematics Studies, Princeton University Press, USA 1987.

D. Rolfsen: *Knots and Links*, American Mathematical Society, USA 1976.

L. W. W. Tu: *An Introduction to Manifolds*, Springer, German 2008.

L. H. Kauffman, I. M. Nikonov, and E. Ogasa: Khovanov–Lipshitz–Sarkar homotopy type for links in thickened higher genus surfaces, *Journal of Knot Theory and Its Ramifications*, 30(8), 2150052 (2021). arxiv 2007.09241 [math.GT].

Boy's paper where Boy discovered Boy surface.

W. Boy: Über die Curvatura integra und die Topologie geschlossener Flächen, *Mathematische Annalen*, 57, 151–184 (1903).

The author cited Boy's paper in the following papers.

E. Ogasa: The projections of n-knots which are not the projection of any unknotted knot, *Journal of Knot Theory and Its Ramifications*, 10, 121–132 (2001), UTMS 97-34, math.GT/0003088.

E. Ogasa: Singularities of projections of n-dimensional knots, *Mathematical Proceedings of Cambridge Philosophical Society*, 126, 511–519 (1999), UTMS96-39, arXiv:1803.03221.

The following paper cited the video "Make Your Boy Surface."

L. H. Kauffman, E. Ogasa, and J. Schneider: A spinning construction for virtual 1-knots and 2-knots, and the fiberwise and welded equivalence of virtual 1-knots, *Journal of Knot Theory and Its Ramifications*, 30(10), 2140003 (2021) Special Issue: Proceedings of 6th Russian — Chinese Conference on Knot Theory and Related Topics (2nd Part).

The author' paper cited in page 149.

E. Ogasa: Ribbon-moves of 2-links preserve the μ-invariant of 2-links, *Journal of Knot Theory and Its Ramifications*, 13(5), 669–687 (2004), math.GT/0004008, UTMS 97-35.

The author's other introductory books are also recommended for readers who are interested in learning more about other topics in four (and higher) dimensions. Note that these books are at a slightly higher level than this book. Two of them are the following.

E. Ogasa: Ijigen e no tobira (In Japanese), Nippon Hyoron Sha Co., Ltd., Japan 2009.

E. Ogasa: *"Seeing Four-Dimensional Space and beyond Using Knots!"* Knots and Everything, World Scientific Pub Co Inc., Singapore 2022.

Index

Epilogue: King of Inventors in the 27th Century

It is in the 27th century.

I was talking in front of the audience.

I showed a Möbius band and a disc. I asked a question, "The boundary of this Möbius band is a single circle. The boundary of this disc is a single circle, too. Can you attach the two circles so that the resulting object does not touch itself?" (This question is the same as Question 7.3 on page 117.)

I gave each of the audience members a piece of *stretchable* paper, a pair of scissors, and a strip of scotch tape. In the 27th century, stretchable paper exists and the price is reasonable. I let them try to do just that.

They had a hard time trying to do that. Several minutes passed.

I tried to move on with my talk and I almost said, "Indeed we cannot attach them like that in our real world, but we can do in four dimensional space..."

Then a girl, one of the audience members, exclaimed, "I shall attach them at any cost!" and she kept on trying.
After a while,

between this person's hands a spinning dark cloud appeared, it lit up and a hole emerged in the space. The girl had opened a door into a high-dimensional space by using a four-dimensional paper construction.